BSA Bantam
Owners Workshop Manual

by Jeff Clew

(117 – 3AG7)

Models covered

Bantam D1. 123cc. '48 to '62
Bantam D3. 148cc. '53 to '57
Bantam D5. 174cc. '57 to '58
Bantam D7. 174cc. '58 to '66
Bantam D10. 174cc. '66 to '67
Bantam D14/4. 174cc. '67 to '68
Bantam de Luxe. 174cc. '64 to '66

Bantam de Luxe. 174cc. '64 to '66
Bantam Sports. 174cc. '66 to '67
Bantam Supreme. 174cc. '66 to '67
Bantam Trials. 123cc. '49 to '55
Bushman. 174cc. '66 to '68
Silver Bantam. 174cc. '65 to '67
Bantam D175. 174cc. '69 to '71

© Haynes Group Limited 2007

ISBN **978 0 85696 117 5**

British Library Cataloguing in Publication Data
A catalogue record for this book is available from the British Library.

Haynes Group Limited
Haynes North America, Inc

www.haynes.com

About this manual

The author of this manual has the conviction that the only way in which a meaningful and easy to follow text can be written is first to do the work himself, under conditions similar to those found in the average household. As a result, the hands seen in the photographs are those of the author. Even the machines are not new; examples that have covered a considerable mileage were selected, so that the conditions encountered would be typical of those found by the average owner/rider. Unless specially mentioned and therefore considered essential, BSA service tools have not been used. There is invariably alternative means of loosening or slackening some vital component, when service tools are not available and risk of damage is to be avoided at all costs.

Each of the six chapters is divided into numbered sections. Within the sections are numbered paragraphs. Cross-reference throughout this manual is quite straightforward and logical. When reference is made, 'See section 6.10' — it means section 6, paragraph 10 in the same chapter. If another chapter were meant it would say 'See chapter 2, section 6.10'.

All photographs are captioned with a section/paragraph number to which they refer, and are always relevant to the chapter text adjacent.

Figure numbers (usually line illustrations) appear in numerical order, within a given chapter. 'Fig.1.1' therefore, refers to the first figure in chapter 1.

Left hand and right hand descriptions of the machines and their components refer to the left and right of a given machine when normally seated facing the front wheel.

Motorcycle manufacturers continually make changes to specifications and recommendations and these, when notified, are incorporated into our manuals at the earliest opportunity.

Whilst every care is taken to ensure that the information in this manual is correct no liability can be accepted by the authors or publishers for loss, damage or injury caused by any errors in or omissions from the information given.

Acknowledgements

Our thanks are due to BSA Motor Cycles Limited for their assistance. Brian Horsfall gave the necessary assistance with the overhaul and devised many ingenious methods for overcoming the lack of service tools. Les Brazier arranged and took the photographs that accompany the text. Tim Parker advised about the way in which the text should be presented and originated the layout.

Our thanks are also due to N. G. Preston Motorcycles of Yeovil, who supplied the "Bantam Super" the model photographed, and to George Jarvis, of Langley Green, Crawley, Sussex who provided much useful advice about the overhaul and repair of these models, based on his widespread experience both as a rider and a repair specialist. We are also indebted to T & G Motorcycles of Thornton Heath for providing technical advice. We are grateful to the Avon Rubber Company who kindly supplied illustrations and advice about tyre fitting, to Amal Limited for their carburettor illustrations, and to John Eyles of Eastleigh for permitting us to photograph the D1 Bantam featured on the front cover of this manual.

Modifications to the BSA Bantam range

The BSA "Bantam" was in production for a total period of 23 years, during which numerous design changes and detail improvements were made. Any significant changes are mentioned in the text, under a separate heading where appropriate. The basic objective is to cover all models, for it is not uncommon to see one of the early rigid frame models still in daily use. It is appreciated that some variants of the models included in this manual were supplied to overseas markets, but in the main these differences relate to the lighting equipment, which has to meet the statutory requirements of the country into which the model is imported.

Contents

Early Bantam D1 model

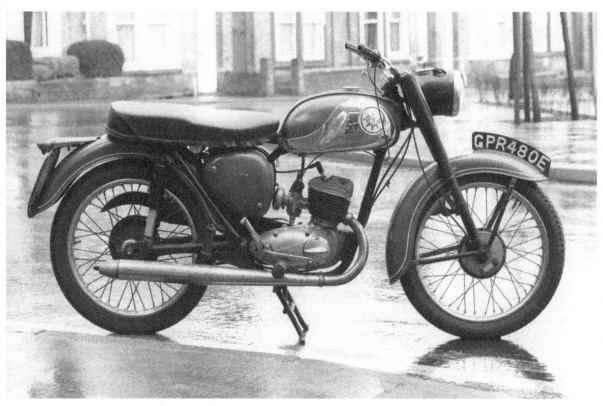

1967 Bantam D14 model

Introduction to the BSA Bantam

During March 1948 BSA Motor Cycles made the surprise announcement that they were to manufacture a batch of 123 cc two-stroke engine/gear units of an entirely new design, as the result of a recently-obtained overseas contract. Previously, the Company had little experience of the two-stroke market, for apart from a brief excursion into this field during the late 1920's, with a 174 cc unit-construction design, they had built their reputation on their well-proven range of four-strokes.

Just three months later, a further announcement gave details of a complete machine, built around the new 123 cc engine/gear unit. Although the initial production was destined for overseas sales, it was apparent from the general reaction to the announcement that a considerable U.K. market existed too. By the end of the year, when the first of the new models was displayed at the Earl's Court Show, there was no longer an embargo on home sales.

The model D1, or "Bantam" as it was named, became immensely popular and was soon a familiar sight on British roads. It was very reliable, had low running costs and was capable of good performance - from what was then considered to be a small capacity engine. Apart from sales to the public, the "Bantam" also found favour with many organisations who required cheap, reliable transport for their messenger fleet. The red-painted "Bantams" on which the GPO Telegram boys made their deliveries became almost as familiar as the London bus. There was also a Competition version of the "Bantam" specially designed for those who indulged in motor cycle trials at weekends. The "Bantam" established a precedent by introducing the hitherto almost unknown entry of a small capacity two-stroke in this type of event.

A 148 cc version, known as the D3 or "Bantam Major" was introduced for the 1954 season, to extend the D1 range. A D3 Competition version was also listed. Some reorganisation of BSA motor cycle production caused the Competition models to be phased out of production during the year following, but this was only a prelude to the intruduction of the 174 cc D5 or "Bantam Super" in 1958, which superceded the D3. The 123 cc and 174 cc models continued in production until 1964, when the former were discontinued. Production of the "Bantam" finally ceased in 1971. Two models were catalogued at this time, a standard road model and a 'trail' riding version, known as the "Bushman". Although no official figures have been published for the total "Bantam" production, BSA claim 150,000 had been sold by February 1957.

Even though production ceased some two years ago, the "Bantam" is anything but extinct. Innumerable racing versions based on the model D1 (and even the ex-GPO machines!) compete regularly in the many short circuit racing events held throughout the UK. Some have been aspired to events of International status, such as the Isle of Man T.T., with no small measure of success either. There is a "Bantam" Racing Club and also "Formula Bantam" events inspired by this Club in which eligible machines must conform to a broad specification that limits the number of modifications permitted. Thus apart from its utilitarian use, the humble "Bantam" has provided many riders with a cheap and convenient means of riding in competitive events without heavy financial outlay or encountering those who are fortunate enough to have sponsors.

One of the more remarkable feats of the "Bantam" that is now almost forgotten is the 14,000 mile journey through Canada, the United States and Mexico by Peggy Iris Thomas, who left England in 1951 with her year-old D1 model, her dog "Matelot" and an amazing amount of luggage and camping equipment. Her adventures on this diminutive machine led to the publication of her book "A Ride in the Sun" and a place in history for one example of what was once Britain's best selling motor cycle.

DIMENSIONS

Wheelbase	50 in.	(127 cm)
Overall length	77.5 in.	(196.9 cm)
Overall height (approx)	36.5 in.	(91.5 cm)
Seat height (unladen)	31 in.	(78.7 cm)
Handlebar width	27.75 in.	(70.49 cm)
Ground clearance	6.75 in.	(17.2 cm)
Unladen weight	215 lbs	(97.5 kg)
Weight of engine/gearbox less carburettor	51 lbs	(23 kg)

(D14 dimensions given here. All other models will be similar)

Ordering spare parts

When ordering spare parts for any of the BSA 'Bantam' range, it is advisable to deal direct with an official BSA agent, who will be able to supply many of the items ex-stock, even though the "Bantam" is no longer in production. Parts cannot be obtained direct from BSA Motor Cycles Limited all orders must be routed through an approved agent, even if the parts required are not held in stock.

Always quote the engine and frame numbers in full, particularly if the parts are required for any of the earlier models. Include any letters before or after the number itself. The frame number will be found stamped at the bottom of the front down tube or along the steering head (early models) and on the front left-hand engine mounting bracket (late models). The engine number is stamped on the left-hand crankcase, to the rear of the cylinder barrel.

Use only parts of genuine BSA manufacture. Pattern parts are available but in many instances they will have an adverse effect on performance and/or reliability. Some complete units are available on a 'service exchange' basis from certain large specialist dealers, (the factory system no longer functions); particularly in London. The usual parts covered are components such as the barrel and crankshaft — it is certainly an economic method of repair. Retain any broken or worn parts until a new replacement has been obtained. Often these parts are required as a pattern for identification purposes, a problem that becomes more acute when a machine is classified as absolute. In an extreme case, it may be possible to reclaim the broken or worn part, or to use it as the pattern for making a new replacement. Many older machines are kept on the road in this way, long after a manufacturer's spares have ceased to be available.

Some of the more expendable parts such as spark plugs, bulbs, tyres, oils and greases etc., can be obtained from accessory shops and motor factors, who have convenient opening hours, charge lower prices and can often be found not far from home. It is also possible to obtain parts on a Mail Order basis from a number of specialists who advertise regularly in the motor cycle magazines.

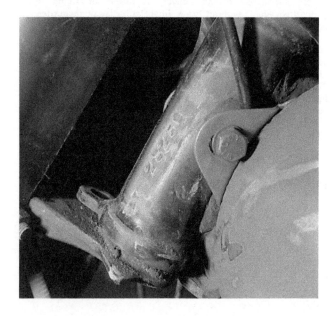

Frame number location - steering head

Engine number location - left hand crankcase

Routine maintenance

Periodic routine maintenance is a continuous process that commences immediately the machine is used. It must be carried out at specified mileage recordings or on a calendar basis if the machine is not used frequently, whichever falls soonest. Maintenance should be regarded as an insurance policy, to help keep the machine in the peak of condition and to ensure long, trouble-free service. It has the additional benefit of giving early warning of any faults that may develop and will act as a regular safety check, to the obvious advantage of both rider and machine alike.

The various maintenance tasks are described under their respective mileage and calendar headings. Accompanying diagrams are provided, where necessary. It should be remembered that the interval between the various maintenance tasks serves only as a guide. As the machine gets older or is used under particularly adverse conditions, it would be advisable to reduce the period between each check.

Some of the tasks are described in detail, where they are not mentioned fully as a routine maintenance item in the text. If a specific item is mentioned but not described in detail, it will be covered fully in the appropriate Chapter. No special tools are required for the normal routine maintenance tasks. The tools contained in the kit supplied with every new machine will prove adequate for each task or if they are not available, the tools found in the average household.

Weekly, or every 200 miles
☐ Oil brake pedal pivot and all exposed control cables and joints.

Monthly, or every 1,000 miles
☐ Check oil level in gearbox and top up if necessary.
☐ Grease the swinging arm pivot (two grease nipples) and the clutch actuating mechanism.
☐ Oil the centre-stand pivots.

Two Months or every 2,000 miles
☐ Change the gearbox oil.
☐ Grease the brake operating arms, taking care not to over-grease and impair braking efficiency by grease reaching the brake linings.
☐ Lubricate final drive chain.

Six Monthly or every 5,000 miles
☐ Grease the rear wheel speedometer drive.
☐ Lubricate the wick that bears on the contact breaker cam, taking great care that oil does not reach the contact breaker points.

Yearly, or every 10,000 miles
☐ Drain and refill the front forks with the recommended grade of oil.
☐ Grease the wheel bearings and the steering head bearings.
☐ These latter two tasks will necessitate a certain amount of dismantling, details of which are given in Chapters 4 and 5.

It should be noted that even when the six monthly and yearly maintenance items have to be undertaken, the weekly and monthly tasks must also be completed. There is no stage, at any point during the life of the machine when a routine maintenance task can be ignored.

Engine	Self-mixing mineral two-stroke oil 20:1 petrol/oil ratio – D1, D3, D5 and D7 24:1 petrol/oil ratio – D10 and later
Gearbox	SAE 40 motor oil 3-speed models - fill to dipstick groove, measured with dipstick fully in place 4-speed models – fill to level screw hole Capacity ¾ pint (425 cc)
Front forks (per leg) D1, D3 and D5 (undamped) D7, D10 and early 　D14/4, D175 (lightweight) Late D14/4 and D175 　(heavyweight)	 Grease SAE 10W/30 ⅛ pt (70 – 75 cc) SAE 10W/30 ⅓ pt (175 cc)
Final drive chain	Aerosol chain lube or motor oil (chain bath for deep lubing)
Spark plug gap	0.018 – 0.020 in
Contact breaker gap	see page 49
Tyre pressures	17 psi (front), 22 psi (rear)
Control cables	Aerosol cable lubricant or light oil
Joints and greasing points	Grease

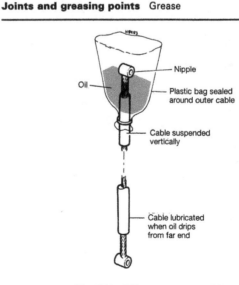

Fig. RM1 Oiling a control cable

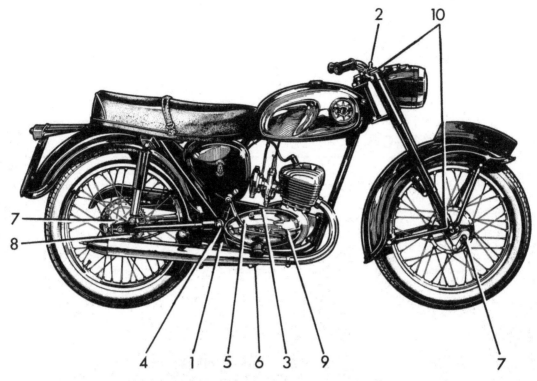

FIG RM2 LUBRICATING POINTS OF THE BSA BANTAM

1 Brake pedal pivot	4 Swinging arm pivots	7 Brake cams	9 Contact breaker cam
2 Control cables	5 Clutch control	8 Speedometer drive cable	10 Front forks
3 Gearbox	6 Centre stud pivots		

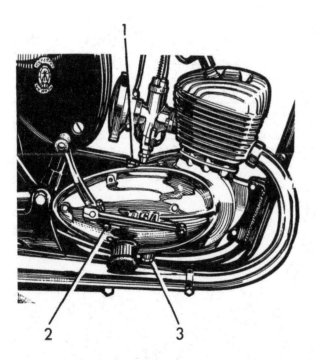

Fig.RM3 Gearbox filler, level and drain plugs

1 Filler plug
2 Oil level screw (4 speed)
3 Drain plug

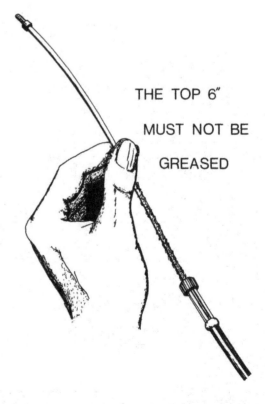

THE TOP 6"

MUST NOT BE

GREASED

Fig.RM4 Speedometer cable greasing

Recommended lubricants

ENGINE	Self-mixing mineral two-stroke oil	Castrol Super TT
GEARBOX	SAE 40 motor oil	Castrol MTX
FINAL DRIVE CHAIN	Aerosol chain lube or motor oil	Castrol Chain Wax
GREASING POINTS	Multi-purpose high melting point lithium based grease	Castrol LM grease

Safety first!

Professional motor mechanics are trained in safe working procedures. However enthusiastic you may be about getting on with the job in hand, do take the time to ensure that your safety is not put at risk. A moment's lack of attention can result in an accident, as can failure to observe certain elementary precautions.

There will always be new ways of having accidents, and the following points do not pretend to be a comprehensive list of all dangers; they are intended rather to make you aware of the risks and to encourage a safety-conscious approach to all work you carry out on your vehicle.

Essential DOs and DON'Ts

DON'T start the engine without first ascertaining that the transmission is in neutral.

DON'T suddenly remove the filler cap from a hot cooling system – cover it with a cloth and release the pressure gradually first, or you may get scalded by escaping coolant.

DON'T attempt to drain oil until you are sure it has cooled sufficiently to avoid scalding you.

DON'T grasp any part of the engine, exhaust or silencer without first ascertaining that it is sufficiently cool to avoid burning you.

DON'T allow brake fluid or antifreeze to contact the machine's paintwork or plastic components.

DON'T syphon toxic liquids such as fuel, brake fluid or antifreeze by mouth, or allow them to remain on your skin.

DON'T inhale dust – it may be injurious to health (see *Asbestos* heading).

DON'T allow any spilt oil or grease to remain on the floor – wipe it up straight away, before someone slips on it.

DON'T use ill-fitting spanners or other tools which may slip and cause injury.

DON'T attempt to lift a heavy component which may be beyond your capability – get assistance.

DON'T rush to finish a job, or take unverified short cuts.

DON'T allow children or animals in or around an unattended vehicle.

DON'T inflate a tyre to a pressure above the recommended maximum. Apart from overstressing the carcase and wheel rim, in extreme cases the tyre may blow off forcibly.

DO ensure that the machine is supported securely at all times. This is especially important when the machine is blocked up to aid wheel or fork removal.

DO take care when attempting to slacken a stubborn nut or bolt. It is generally better to pull on a spanner, rather than push, so that if slippage occurs you fall away from the machine rather than on to it.

DO wear eye protection when using power tools such as drill, sander, bench grinder etc.

DO use a barrier cream on your hands prior to undertaking dirty jobs – it will protect your skin from infection as well as making the dirt easier to remove afterwards; but make sure your hands aren't left slippery. Note that long-term contact with used engine oil can be a health hazard.

DO keep loose clothing (cuffs, tie etc) and long hair well out of the way of moving mechanical parts.

DO remove rings, wristwatch etc, before working on the vehicle – especially the electrical system.

DO keep your work area tidy – it is only too easy to fall over articles left lying around.

DO exercise caution when compressing springs for removal or installation. Ensure that the tension is applied and released in a controlled manner, using suitable tools which preclude the possibility of the spring escaping violently.

DO ensure that any lifting tackle used has a safe working load rating adequate for the job.

DO get someone to check periodically that all is well, when working alone on the vehicle.

DO carry out work in a logical sequence and check that everything is correctly assembled and tightened afterwards.

DO remember that your vehicle's safety affects that of yourself and others. If in doubt on any point, get specialist advice.

IF, in spite of following these precautions, you are unfortunate enough to injure yourself, seek medical attention as soon as possible.

Asbestos

Certain friction, insulating, sealing, and other products – such as brake linings, clutch linings, gaskets, etc – contain asbestos. *Extreme care must be taken to avoid inhalation of dust from such products since it is hazardous to health.* If in doubt, assume that they *do* contain asbestos.

Fire

Remember at all times that petrol (gasoline) is highly flammable. Never smoke, or have any kind of naked flame around, when working on the vehicle. But the risk does not end there – a spark caused by an electrical short-circuit, by two metal surfaces contacting each other, by careless use of tools, or even by static electricity built up in your body under certain conditions, can ignite petrol vapour, which in a confined space is highly explosive.

Always disconnect the battery earth (ground) terminal before working on any part of the fuel or electrical system, and never risk spilling fuel on to a hot engine or exhaust.

It is recommended that a fire extinguisher of a type suitable for fuel and electrical fires is kept handy in the garage or workplace at all times. Never try to extinguish a fuel or electrical fire with water.

Note: *Any reference to a 'torch' appearing in this manual should always be taken to mean a hand-held battery-operated electric lamp or flashlight. It does **not** mean a welding/gas torch or blowlamp.*

Fumes

Certain fumes are highly toxic and can quickly cause unconsciousness and even death if inhaled to any extent. Petrol (gasoline) vapour comes into this category, as do the vapours from certain solvents such as trichloroethylene. Any draining or pouring of such volatile fluids should be done in a well ventilated area.

When using cleaning fluids and solvents, read the instructions carefully. Never use materials from unmarked containers – they may give off poisonous vapours.

Never run the engine of a motor vehicle in an enclosed space such as a garage. Exhaust fumes contain carbon monoxide which is extremely poisonous; if you need to run the engine, always do so in the open air or at least have the rear of the vehicle outside the workplace.

The battery

Never cause a spark, or allow a naked light, near the vehicle's battery. It will normally be giving off a certain amount of hydrogen gas, which is highly explosive.

Always disconnect the battery earth (ground) terminal before working on the fuel or electrical systems.

If possible, loosen the filler plugs or cover when charging the battery from an external source. Do not charge at an excessive rate or the battery may burst.

Take care when topping up and when carrying the battery. The acid electrolyte, even when diluted, is very corrosive and should not be allowed to contact the eyes or skin.

If you ever need to prepare electrolyte yourself, always add the acid slowly to the water, and never the other way round. Protect against splashes by wearing rubber gloves and goggles.

Mains electricity and electrical equipment

When using an electric power tool, inspection light etc, always ensure that the appliance is correctly connected to its plug and that, where necessary, it is properly earthed (grounded). Do not use such appliances in damp conditions and, again, beware of creating a spark or applying excessive heat in the vicinity of fuel or fuel vapour. Also ensure that the appliances meet the relevant national safety standards.

Ignition HT voltage

A severe electric shock can result from touching certain parts of the ignition system, such as the HT leads, when the engine is running or being cranked, particularly if components are damp or the insulation is defective. Where an electronic ignition system is fitted, the HT voltage is much higher and could prove fatal.

Chapter 1 Engine, Clutch and Gearbox unit

Contents

Specifications

The BSA 'Bantam' models all employ the same basic engine/gear unit in which the gearbox is an integral part of the engine assembly. The gearbox has either three or four gears, depending on the year of manufacture of the machine and the general specifcation of the model. Although the information given relates specifically to the 174 cc models, it applies equally well to the earlier 148 cc and 123 cc models, which use a similar engine/gear unit with reduced cylinder bore dimensions. These models have a three-speed gearbox only.

The same dismantling and reassembly procedure is applicable to ALL 'Bantam' models. Where any significant changes in design have been made, mention is included in the text, together with the revised procedure necessary.

The Chapter concludes with a fault finding table.

Engine

Type ...	...	...	...	...	...	...	...	...	Single cylinder two-stroke, with petrol lubrication
Cylinder head..	...	...	...	...	...	...	...	...	Cast iron (D1 models, 1948–53)
									Cast aluminium alloy (Later D1 and all other models)
Cylinder barrel	...	...	...	...	...	...	...	...	Cast iron

Bore	61.5 mm (D5, D7, D10 and D14 models)	
	57 mm (D3 model)	
	52 mm (D1 model)	
Stroke	58 mm all models	
Capacity	174 cc (D5, D7, D10 and D14 models)	
	148 cc (D3 model)	
	123 cc (D1 model)	
bhp	13 @ 5,750 rpm (D14 models)	
	10 @ 6,000 rpm (D10 models)	
	7.5 @ 5,000 rpm (D5 and D7 models)	
	4.9 @ 4,750 rpm (D3 model)	
	4 @ 5,000 rpm (D1 model)	
Compression ratio	10 : 1 (D14 models)	
	8.65 : 1 (D10 models)	
	7.4 : 1 (D5 and D7 models)	
	6.4 : 1 (D3 model)	
	6.5 : 1 (D1 model)	

Piston

Clearance (top of skirt)	0.0077 inch - 0.0078 inch
Clearance (bottom of skirt)...	0.0038 inch - 0.0039 inch
Oversizes available	+0.020 inch and +0.040 inch

Piston Rings

Compression (two rings)	Ends profiled to engage with piston pegs
Radial depth	0.078 inch
Width	0.062 inch
End gap	0.009 inch - 0.013 inch

Ports

Inlet port size	1 inch
Exhaust port size	1.25 inch

Torque Wrench Settings

Cylinder head nuts	18 - 20 ft/lb
Carburettor stud nuts	10 - 12 ft/lb

Recommendations for D14 models
Gear Ratios

D1 models

	Standard.	Competition
Top	7.0 : 1	8.64 : 1
Second	11.7 : 1	14.45 : 1
Bottom	22.0 : 1	27.1 : 1

D3 models.. ... ...   As D1 models above

D5 and D7 models (early)

Top	6.48 : 1
Second..	10.74 : 1
Bottom	20.2 : 1

D7 models (late)

Top	6.58 : 1
Second	9.26 : 1
Bottom	17.4 : 1

D10 and D14 models

Top	6.58 : 1
Third	8.55 : 1
Second	12.04 : 1
Bottom	18.68 : 1

'Bushman' models

Top	8.10 : 1
Third	10.54 : 1
Second	14.83 : 1
Bottom	23.03 : 1

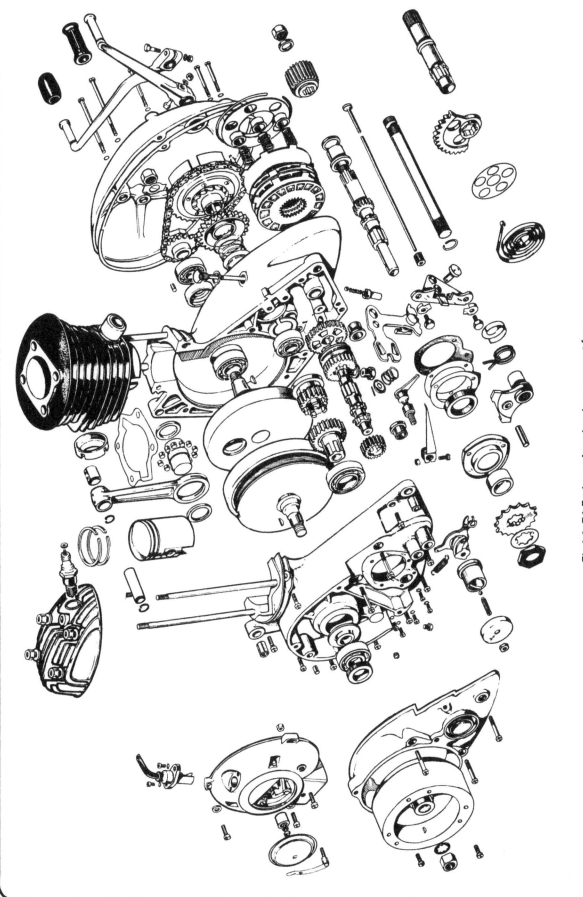

Fig.1.1. D1 Engine and gearbox in component form

Sprocket Sizes

							Engine	Clutch	Gearbox	Rear wheel
D1 models (standard)	...	...	...	...	...	...	17	38	15	47
D1 models (competition)	...	...	...	...	...	...	17	38	15	58
D3 models (standard)	...	...	...	...	...	...	17	38	15	47
D3 models (competition)	...	...	...	...	...	...	17	38	15	58
D5 and D7 models (early)	...	...	...	...	...	...	17	38	16	46
D7 models (late)	...	...	...	...	...	...	17	38	16	47
D10 and D14 models	...	...	...	...	...	...	17	38	16	47
'Bushman' models	...	...	...	...	...	...	17	38	16	58

Chain Sizes

Primary	...	...	...	...	...	...	...	...	3/8 inch x .250 inch
Final drive	...	...	...	...	...	...	...	...	½ inch x .205 inch

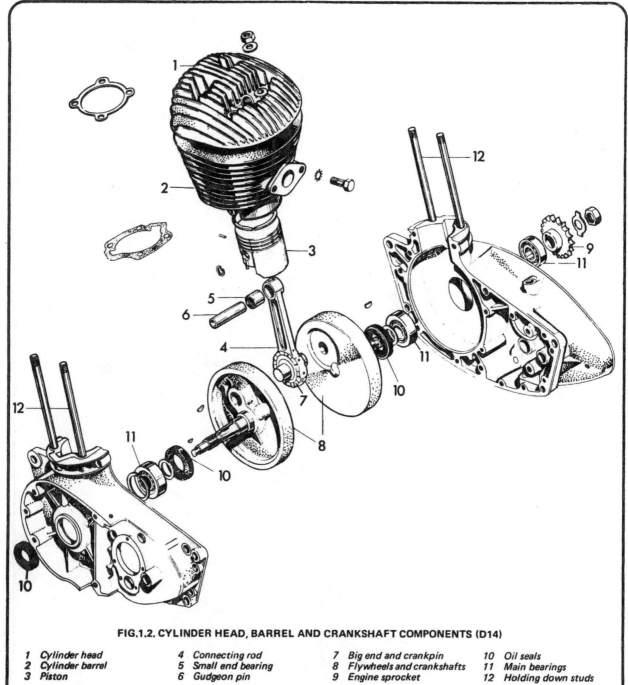

FIG.1.2. CYLINDER HEAD, BARREL AND CRANKSHAFT COMPONENTS (D14)

1 Cylinder head	*4 Connecting rod*	*7 Big end and crankpin*	*10 Oil seals*
2 Cylinder barrel	*5 Small end bearing*	*8 Flywheels and crankshafts*	*11 Main bearings*
3 Piston	*6 Gudgeon pin*	*9 Engine sprocket*	*12 Holding down studs*

1 General Description

The engine fitted to the BSA 'Bantam' models is of the single cylinder two-stroke type, using a flat top piston and what is known as 'loop scavenging' to effect a satisfactory induction and exhaust sequence. The later D14 models have enlarged ports and thickly padded flywheels to increase crankcase compression. The piston rings are pegged in characteristic two-stroke practice, to prevent the rings from rotating and becoming trapped in the cylinder barrel ports. Large diameter oil seals form an effective crankcase seal around the crankshaft, which runs on three journal ball bearings, two on the drive side and one on the generator side. Endfloat is controlled by shims. All engine/gear castings are in aluminium alloy; the same material is employed for the cylinder head. Only the cylinder barrel is of cast iron.

The flywheel generator is on the left hand side of the engine. It is of the a.c. alternator type, with the rotor attached to the end of the crankshaft. The clutch assembly is located on the right hand side of the machine, behind a pear-shaped aluminium cover. It is necessary to dismantle the clutch before the crankcases can be separated. The exhaust system is carried on the right hand side of the machine, in a downswept layout. A kickstarter and a gear change lever on the right hand side of the machine share the same common pivot.

Lubrication is effected by the petroil system, in which a measured amount of self-mixing oil is dissolved in the petrol. Because a two-stroke engine relies on crankcase compression as part of its mode of operation, the oil content of the petrol is distributed to the crankshaft, main bearings and big-end assembly, as well as to all other working parts of the engine. This system works on the total loss principle, all excess oil being discharged via the exhaust. In consequence there is no necessity for an oil change; every incoming charge brings with it a fresh quota of clean oil. There is no significant build-up of oil in the crankcase; the gearbox has its own separate oil content.

2 Operations with Engine in Frame

It is not necessary to remove the engine unit from the frame unless the crankshaft assembly and/or the gearbox internals require attention. Most operations can be accomplished with the engine in place, such as:
1 Removal and replacement of cylinder head.
2 Removal and replacement of cylinder barrel and piston.
3 Removal and replacement of flywheel generator.
4 Removal and replacement of clutch assembly.
5 Removal and replacement of contact breaker assembly.

When several operations need to be undertaken simultaneously, it will probably be advantageous to remove the complete engine unit from the frame, an operation that should take approximately fifteen minutes. This will give the advantage of better access and more working space.

3 Operations with Engine Removed

1 Removal and replacement of the main bearings.
2 Removal and replacement of the crankshaft assembly.
3 Removal and replacement of the gear cluster, selectors and gearbox main bearings.
4 Replacement of the gear lever return spring.

4 Method of Engine/Gearbox Removal

As described previously, the engine and gearbox are built in unit and it is necessary to remove the unit complete, in order to gain access to either component. Separation is accomplished after the engine unit has been removed and refitting cannot take place until the crankcases have been reassembled. When the crankcases are separated the gearbox internals will also be exposed.

5 Removing the Engine/Gear Unit

1 Place the machine on the centre stand and make sure it is standing firmly, on level ground.
2 Turn off the fuel supply and disconnect the fuel pipe where it joins the carburettor float chamber by disconnecting the union nut. Disconnect also the air cleaner hose at the carburettor intake (if any). Slacken and remove the two nuts and washers securing the carburettor to the cylinder flange studs (or slacken clip retaining carburettor to intake stub - early models) and tie carburettor out of way. Remove the 'O' ring (if fitted).
3 Use the 'C' spanner in the tool kit to unscrew the slotted exhaust pipe union nut from the cylinder barrel. This may prove difficult to start, especially if the nut has not been disturbed for some while. Application of a penetrating oil such as 'Plus Gas' will help, if the oil is applied liberally and left to soak into the joint for a few seconds. Do not use force, otherwise the thread may be stripped or worse still, the stub cracked or broken.
4 Remove the nuts and bolts attaching the silencer to the frame attachment point and withdraw the exhaust system complete.
5 Detach the sparking plug lead and disconnect the generator leads by means of the snap connectors. Note the wires are colour coded, to make reconnection easy.
6 Disconnect the clutch cable at the handlebar end and coil it so that the cable does not impede the dismantling operation.
7 Remove the chainguard and the final drive chain by detaching the spring link.
8 The engine is held in the frame by two bolts and nuts through the front engine bracket and by two bolts at the rear. Remove all four bolts and lift the engine unit from the frame.
9 On the later models it will also be necessary to remove the two bolts at the petrol tank front mounting and the single bolt at the rear of the tank. This will permit the tank to be raised slightly, to provide the clearance necessary during removal of the engine.
10 On the trials models the front engine bracket bolts retain also the crankcase undershield.
11 Remove the kickstarter and gear change pedals, after slackening the pinch bolts that retain them on their splined shafts.

6 Dismantling the Engine, Clutch and Gearbox - General

Before commencing work on the engine unit, the external surfaces should be cleaned thoroughly. A motor cycle engine has very little protection from road grit and other foreign matter, which will find its way into the dismantled engine if this simple precaution is not observed. One of the proprietary cleaning compounds such as 'Gunk' can be used to good effect, particularly if the compound is allowed to work into the film of oil and grease before it is washed away. When washing down, make sure that water cannot enter the carburettor or the electrical system, particularly if these parts have been exposed.

Never use undue force to remove any stubborn part, unless mention is made of this requirement. There is invariably good reason why a part is difficult to remove, often because the dismantling operation has been tackled in the wrong sequence. Dismantling will be made easier if a simple engine stand is constructed that will correspond with the engine mounting points. This arrangement will permit the complete unit to be clamped rigidly to the work bench, leaving both hands free.

7 Dismantling the Engine, Clutch and Gearbox - Removing the Cylinder Head, Barrel and Piston

1 Remove the four ¼ inch nuts that retain the cylinder head in position and lift the cylinder head clear of the retaining studs.
2 Remove the cylinder barrel by first slackening the two screws at the cylinder base flange (left-hand side). Slide the barrel upwards along the retaining studs, taking care to catch the piston as it emerges from the cylinder bore. If this precaution is not

observed, there is risk of damage to the piston and/or piston rings.

3 There should be sufficient clearance to lift the cylinder barrel clear of the retaining studs on all but the very late models, so that it can be put aside for further attention later. On the D10 and D14 models, it will be necessary to unscrew the two front fixing bolts of the petrol tank and also to loosen the rear fixing bolt, so that the tank can be raised sufficiently to give enough clearance.

4 Remove the piston complete with rings by detaching one of the gudgeon pin circlips with a pair of long nose pliers. If only a 'top' overhaul is contemplated at this time, it is advisable to pad the mouth of the crankcase with rag, to prevent the mishap of a displaced circlip falling in.

5 If the gudgeon pin is a tight fit and cannot be withdrawn easily, warm the piston with a rag soaked in hot water. This will expand the piston bosses sufficiently to release their hold on the pin. Before the pin is withdrawn, make sure the crankcase mouth is covered, to obviate the risk of broken piston rings falling in. Remove the piston complete.

6 Mark the piston INSIDE the skirt so that it is replaced in the same position. It is essential that the piston ring gaps, as denoted by the ring pegs, face the FRONT of the engine, otherwise the rings will be trapped in the ports and broken.

8 Clutch - Dismantling and Removal

1 Remove the primary drive cover of the right-hand crankcase, which is held in place by five screws. There is no need to remove the oil level screw painted red, that is fitted to the later models.

2 Place a container below the cover to catch the oil that will be released and break the joint by tapping lightly with a rawhide mallet. Lift away the cover which, in the case of the D10 and D14 models, will contain also the contact breaker plate assembly. There will be either one or two locating dowels, depending on the model.

3 Detach the primary chain by removing the spring link. Take off the clutch cover plate, retained by three small screws, each with a spring washer.

4 With the aid of a second pair of hands, compress the clutch pressure plate sufficiently to release the large circlip that holds the clutch assembly together. It is located just inside the clutch drum. Remove the clutch springs and caps (six), the clutch plates (three inserted, three plain) and the mushroom-headed push rod that fits within the gearbox mainshaft.

5 Unscrew the centre retaining nut of the clutch drum and pull the drum off the gearbox mainshaft. It will be necessary to lock the clutch sprocket before sufficient force can be applied to slacken the nut (right-hand thread). The clutch centre, clutch drum, kickstarter ratchet and centre bush will become free, when this nut is slackened and removed. If desired, it is possible to leave these parts in position until a later stage of the dismantling sequence.

9 Kickstarter Shaft and Quadrant - Removal

1 When the complete clutch assembly has been removed, the kickstarter shaft can be pulled from its housing, complete with the quadrant, clock-type return spring and perforated washer.

2 Note how tension is applied to the return spring by looping one end of the spring over a projection in the crankcase casting. Care is necessary when removing the spring because it will be under tension.

10 Engine Sprocket - Removal

1 Before the engine sprocket can be removed on the D10 and D14 models, it is first necessary to withdraw the contact breaker cam. Unscrew the contact breaker cam screw, which will be found in the centre of the cam. Before the screw reaches the end of the thread, it will encounter a circlip that is fitted to prevent

5.3. Toolkit 'C' spanner fits exhaust pipe nut

5.4. Silencer is clamped to frame by captive bolts

5.7. Disconnection is easiest at rear sprocket

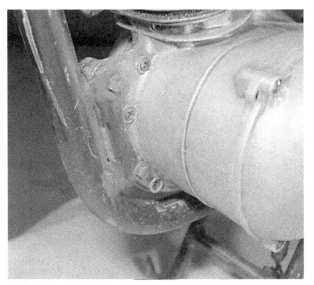

5.8A. Front engine bolts

5.8B. Rear engine bolts

5.11. Pinch bolts retain levers on splines

6. Clean exterior of engine unit prior to dismantling

7.6. Piston ring pegs should face FRONT of engine

8.1. Removal of right hand cover exposes primary drive

8.3A. Disconnection is easiest at clutch sprocket

8.3B. Do not lose spring washers when removing screws

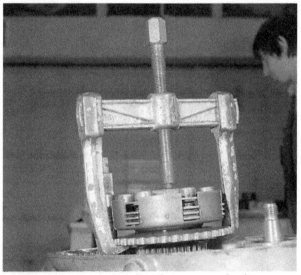

8.4. Sprocket puller can be adapted to compress clutch

8.5. Clutch centre lifts off mainshaft

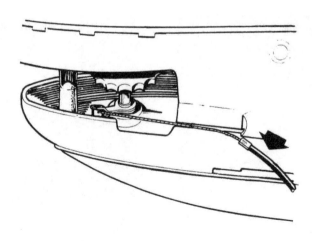

Fig.1.3. Clutch cable removal

the screw from working free. Further rotation of the screw will have an extractor effect, drawing the cam from the crankshaft.

2 It may be necessary to tap the cam very gently, to free it from its taper.

3 Remove the nut securing the engine sprocket in position (self-locking nut, D10 and D14 models). To prevent the crankshaft from turning, pass a stout metal rod through the small end of the connecting rod. The sprocket is keyed on to the crankshaft; it may be necessary to use a sprocket puller to avoid damaging the thread on the end of the crankshaft.

4 When the sprocket has been removed, withdraw the woodruff key from the crankshaft, also the collar and oil seal in front of the crankshaft main bearing.

11 Generator - Removal

D10 and D14 models only

1 Reverting to the left-hand side of the engine, remove the four screws that secure the generator cover to the left-hand crankcase. Withdraw the cover, taking note that a gasket is not needed at this joint.

2 Remove the inner cover by unscrewing the cross head screw to the rear of the clutch adjuster. Unscrew the three generator stator nuts, noting that the forward-mounted nut is extended to accept one of the outer cover screws.

3 Tap the cover gently to break the joint and remove the cover complete with the stator coil assembly. Remove and put aside the small spacers on the fixing studs, located between the stator and the inner cover. The inner cover can now be pulled free.

D1, D3, D5 and D7 models only

4 A slightly different approach is necessary because the contact breaker assembly is mounted on the left-hand side of the engine, immediately in front of the flywheel generator. Commence by removing the outer screws that retain the stator plate assembly in position and withdraw the screw from the centre of the contact breaker cam. The cam, which is keyed on to the crankshaft, can now be withdrawn and the stator plate removed complete after the cable harness has been threaded back through the hole in the generator cover. Take care not to lose the small key that locates the contact breaker cam in its correct position.

5 The generator rotor is keyed to the crankshaft and retained by a large centre nut (right hand thread) and spring washer. Lock the crankshaft by passing a stout metal rod through the small end of the connecting rod and unscrew the nut. The rotor can now be pulled off the crankshaft preferably by using BSA Service Tool 61-3188, but it can sometimes be removed by giving a few gently taps around the inner rim, to help break the taper joint. This is a very delicate operation, without the Service Tool; the crankshaft will bend unless great care is taken. When the rotor has been removed, withdraw the Woodruff key from the crankshaft and put it aside until reassembly commences.

6 When withdrawn, the flywheel rotor must be placed in its correct position within the stator coil assembly, or covered with a circular steel plate. Failure to observe these instructions may entail loss of electrical efficiency because the magnetic properties of the flywheel are not retained.

7 The inner cover can now be removed by unscrewing the five retaining screws, two of which are located behind the flywheel rotor (Wipac ignition models). Only three screws retain the cover when Lucas ignition equipment is employed.

8 The inner cover will pull away complete with the clutch operating mechanism (all models). Access is now available to the gearbox final drive sprocket and the gear indicator lever (early models only).

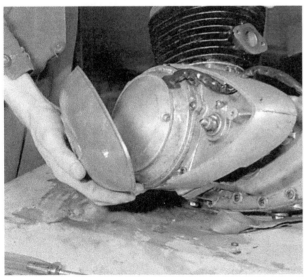

11.4A. Removal of outer cover gives access to generator

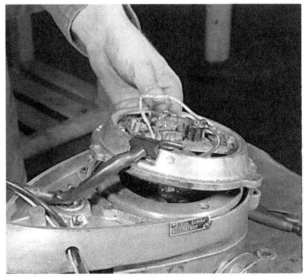

11.4B. Stator plate can be lifted off with wiring harness

12 Gearbox Final Drive Sprocket - Removal

1 The gearbox final drive sprocket is secured to the end of the gearbox mainshaft by a large hexagon nut and tab washer. Bend back the tab washer and unscrew the nut, taking note that it has a LEFT-HAND thread.

2 The sprocket can be held firmly during this operation by wrapping the chain around it and clamping both ends of the chain in a vice. If a vice is not available, the sprocket can be held by wrapping the chain as shown in the accompanying diagram.

13 Separating the Crankcases

1 The crankcases can be separated by unscrewing the 16 screws that clamp both halves together. 15 of these screws are on the left-hand side of the engine unit; one is found within the outside right-hand crankcase casting above the gearbox main bearing. Do not overlook the two screws at the cylinder base flange. Early models have only 11 screws.

2 Note that two locating dowels are fitted, in each of the uppermost engine bolt holes.

3 Never lever the crankcases apart with a screwdriver because the efficiency of any two-stroke engine is dependent on a leak proof crankcase joint. The only safe method of separation is to tap the cylinder barrel studs with a rawhide mallet.

14 Crankshaft Assembly - Removal

1 The crankshaft assembly will most probably be retained in the right hand crankcase, where the crankshaft passes through two journal ball bearings. Irrespective of which crankcase holds it captive, a light tap on the end of the crankshaft with a rawhide mallet should be sufficient to part the assembly from the main bearing(s).

2 Make a careful note of the number and thicknesses of any shims fitted to either side of the flywheel assembly. When new, two 0.010 inch shims are usually fitted, to give the desired amount of end float to the crankshaft assembly.

15 Gearbox Components - Removal

D14 and other four-speed gearbox models

1 Commence by removing the loose fitting spring and the claw-shaped selector from the gear change spindle, then withdraw the spindle complete with spring and spacer.

2 Withdraw the layshaft, complete with fixed top and second gears. The remaining layshaft gears can now be displaced, noting that the sliding third gear is positioned with its pegs facing bottom gear. Note also that the layshaft second gear is held against the fixed top gear by a circlip.

3 Before the mainshaft gears can be removed, it is necessary to take off the cam plate mounting bracket. Bend back the two tab washers and take out the two bolts. The cam plate can now be lifted clear, complete with the two selector forks, and also the sliding second gear of the mainshaft.

4 Take care not to misplace the two loose fitting fork rollers when disengaging the selector forks from the cam plate tracks.

5 If the quadrant is separated from the cam plate note how these two components are assembled. Correct indexing of the gears depends on their correct alignment.

6 When the cam plate assembly has been removed, the cam plate ball, seating and spring can be withdrawn from the crankcase. It is unlikely that the plunger socket will need to be removed, or the selector fork spindle. The former is a press fit and can be removed by tapping with a drift from the outside. The latter is held in position by a circlip.

7 If the clutch centre and drum have not already been withdrawn, clamp the gearbox mainshaft in a vice fitted with soft jaw clamps. This will permit the centre retaining nut to be unscrewed (right hand thread) and the clutch centre, clutch drum, kickstarter ratchet and centre bush to be removed.

8 Withdraw the gearbox mainshaft, complete with the fixed bottom gear. Note that third gear is retained against bottom gear by a circlip. The clutch operating push rod is located within the hollow mainshaft.

12.1. This nut has a left hand thread

13.2. Note position of crankcase locating dowels

15.10. Three-speed models have the distinctive selector claw

All three-speed gearbox models

9 Although the earlier models have only a three-speed gearbox, the general layout and operating principle is virtually identical with regard to the D5, D7 and D10 models. The instructions for dismantling the four-speed gearbox can therefore be followed in broad terms, without need for any special instructions.

10 The D1 and D3 models employ a somewhat similar arrangement, with a different form of gear selector mechanism. The selector takes the form of a claw which is held in position by a circlip. The boss of the claw is surrounded by a double-ended coil spring, housed within a metal cover. The two ends of the spring fit either side of a peg driven into the claw and act as a means of centralising the claw during gear changing operations. There is no cam plate used in this form of assembly; the claw arrangement fulfills the indexing function.

16 Crankshaft and Gearbox Main Bearings - Removal

1 The crankshaft assembly runs on three journal ball bearings, two on the drive side and one on the generator side (right-hand and left-hand sides respectively). Before any of these bearings can be removed, the oil seals must first be prised from their housings.

2 Before the bearings can be removed, it is necessary to warm the crankcases by applying a rag soaked in hot water. The inner bearing on the drive side and the bearing on the generator side are retained in position by circlips (D10 and D14 models) and will press to the inside of the crankcases. The smaller outer race on the drive side will press to the outside of the crankcase. Early models have the same bearing arrangement, without the circlips.

3 The gearbox mainshaft runs on a journal ball bearing in the left-hand crankcase and a roller bearing in the right-hand crankcase (D10 and D14 models only). Each is retained by a circlip and is driven inwards when replacement is necessary, after the crankcase has been warmed. The earlier three-speed models (D1, D3, D5 and D7) employ two journal ball bearings. The bearing in the left-hand crankcase is covered by a metal oil seal housing located by three 3/16 inch screws and washers which, when removed, reveals a metal plate held by two 3/16 inch screws and washers. This latter plate serves the dual function of locating both the main bearing for the mainshaft and the phosphor bronze bush for the layshaft.

4 The two journal ball bearings and the two phospor bronze layshaft bushes (one in each crankcase) will press to the inside when the crankcases have been warmed.

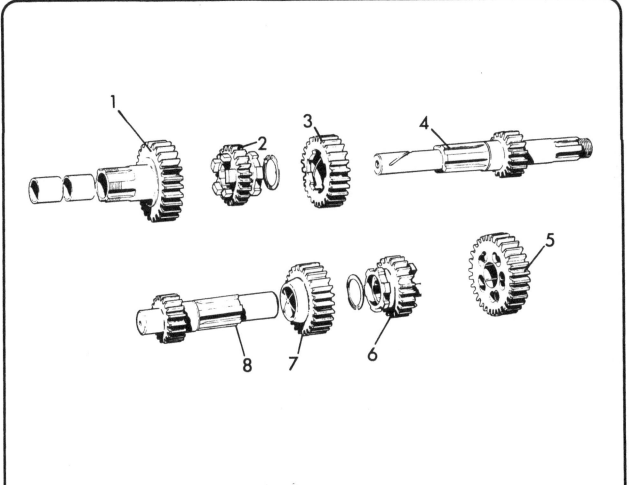

Fig.1.4. Gear cluster (4 speed)

1 Top gear (mainshaft)	*3 3rd gear (mainshaft)*	*5 1st gear (layshaft)*	*7 2nd gear (layshaft)*
2 2nd gear (mainshaft)	*4 Mainshaft with integral 1st gear*	*6 3rd gear (layshaft)*	*8 Layshaft with integral top gear*

17 Main Bearings and Oil Seals - Examination

1 When the bearings have been pressed from their housings, wash them with a petrol/paraffin mix, to remove all traces of oil. If there is any play in the ball or roller bearings, or if they do not revolve smoothly, new replacements should be fitted. With regard to the phosphor bronze bushes, any signs of wear or roughness will immediately be evident. Replace them if there is any doubt about their condition.

2 It is highly desirable to replace the oil seals particularly those fitted to the crankshaft. Worn oil seals will admit air to the crankcase of a two-stroke engine, which will dilute the incoming mixture whilst it is under crankcase compression. Crankcase air leaks are the most frequent cause of difficult starting and uneven running in any two-stroke engine.

3 Do not overlook the phosphor bronze bearing in the centre of the flywheel magneto stator plate assembly. When wear occurs, crankshaft whip will cause a constantly varying contact breaker gap. The bush is a tight push fit in its housing.

18 Crankshaft Assembly - Examination and Renovation

1 Wash the complete flywheel assembly with a petrol/paraffin mix to remove all surplus oil. Then hold the connecting rod at its highest point of travel (fully extended) and check whether there is any vertical play in the big-end bearing by alternately pulling and pushing in the direction of travel. If the bearing is sound, there should be no play whatsoever.

2 Ignore any sideplay unless this appears to be excessive. A certain amount of play in this direction is necessary if the bottom end of the engine is to run freely.

3 Although it may be possible to run the engine for a further short period of service with a very small amount of play in the big-end bearing, this course of action is not advisable. Apart from the danger of the connecting rod breaking if the amount of wear increases rapidly, a further complete engine strip will be necessary to effect the renewal. It is best to replace the big-end bearing at this stage, if it is in any way suspect. Wear is denoted by the characteristic 'knock' when the engine is running under load.

4 A Works reconditioned flywheel assembly was available originally as a replacement, but because the "Bantam" is no longer in production only a few specialist BSA agents can provide this service. The alternative is to separate the flywheels and fit a new big end bearing, a highly specialised task that requires a press for separating the flywheels and a lathe for re-aligning the rebuilt assembly. This work is best entrusted to a qualified repairer, especially since the average owner will not have the necessary experience or access to the equipment needed.

19 Cylinder Barrel - Examination and Renovation

1 There will probably be a lip at the uppermost end of the cylinder barrel, which marks the limit of travel of the top piston ring. The depth of the lip will give some indication of the amount of bore wear that has taken place, even though the amount of wear is not evenly distributed.

2 Remove the rings from the piston taking great care as they are brittle and very easily broken. There is more tendency for the rings to gum in their grooves in a two-stroke engine. Insert the piston in the bore so that it is positioned about ¾ inch below the top of the bore. If it is possible to insert a 0.012 inch feeler gauge between the piston and the bore, a rebore and the fitting of an oversize piston is necessary.

3 Give the cylinder barrel a close visual inspection. If the surface of the bore is scored or grooved, indicative of an earlier seizure or a displaced circlip and gudgeon pin, a rebore is essential. Compression loss will have a very marked effect on performance.

4 Check that the outside of the cylinder barrel is clean and free from road dirt. Use a wire brush on the cooling fins if they are obstructed in any way. The engine will overheat badly if the cooling area is obstructed in any way. The application of matt cylinder black will help improve heat radiation.

5 Clean all carbon deposits from the exhaust ports and try and obtain a smooth finish in the ports without in any way enlarging them or altering their shape. The size and position of the ports predetermines the characteristics of the engine and unwarranted tampering can produce very adverse effects. An enlarged or re-profiled port does not necessarily guarantee an increase in performance.

20 Piston and Piston Rings - Examination and Renovation

1 Attention to the piston and piston rings can be overlooked if a rebore is necessary because new replacements will be fitted.

2 If a rebore is not considered necessary, the piston should be examined closely. Reject the piston if it is badly scored or if it is badly discoloured as the result of the exhaust gases by-passing the rings.

3 Remove all carbon from the piston crown and use metal polish to finish off. Carbon will not adhere so readily to a polished surface.

4 Check that the gudgeon pin bosses are not worn or the circlip grooves damaged. Check also the piston ring pegs, to make sure that none has worked loose.

5 The grooves in which the piston rings locate can also become enlarged in use. The clearance between the piston and the ring in the groove should not exceed 0.002 inch.

6 Piston ring wear can be checked by inserting the rings in the

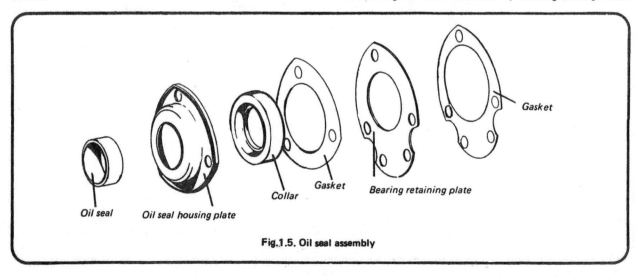

Oil seal   Oil seal housing plate   Collar   Gasket   Bearing retaining plate   Gasket

Fig.1.5. Oil seal assembly

cylinder bore from the top and pushing them down about 1½ inches with the crown of the piston so that they rest square in the cylinder. If the end gap exceeds 0.013 inch the rings should be replaced.

7 Examine the working surface of the rings. If discoloured areas are evident, the rings should be replaced since the patches indicate the blow-by of gas. Check also that there is not a build-up of carbon behind the tapered ends of the rings, where they locate with the piston ring pegs.

8 It cannot be over-emphasised that the condition of the piston and rings in a two-stroke engine is of prime importance, especially since they control the opening and closing of the ports in the cylinder barrel by providing an effective seal. A two-stroke engine has only three working parts, one of which is the piston. It follows that the efficiency of the engine is very dependent on condition of this component and the parts with which it is closely associated.

21 Small-End - Examination and Renovation

1 The small end bearing of a two-stroke engine is more prone to wear, producing the characteristic rattle that is heard in many engines that have covered a considerable mileage. The gudgeon pin should be a good sliding fit in the bearing, without evidence of any play. If play is apparent, the bearing must be renewed.

2 The D10 and D14 models employ a bush containing needle roller bearings. Earlier models have a plain phosphor bronze bush. In either case the new bearing can be used to push the old bearing out of position by using a drawbolt and distance piece arrangement as shown in the accompanying diagram.

3 When the new bush is in position, it will be necessary to ream it out to the correct size, to offset the slight compression that has occurred when the bush was drawn into position. Only a very small amount of metal will need removing, otherwise the gudgeon pin fit will be too slack.

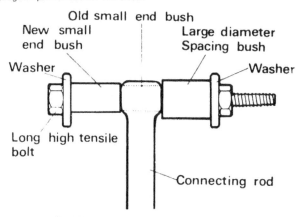

20.2 Method of inserting small end bush

22 Cylinder Head - Examination and Renovation

1 It is unlikely that the cylinder head will require any special attention apart from removing the carbon deposit from the combustion chamber. Finish off with metal polish; a polished surface will reduce the tendency for carbon to adhere and will also help improve the gas flow.

2 Check that the cooling fins are not obstructed so that they receive the full air flow. A wire brush provides the best means of cleaning.

3 Check the condition of the thread where the sparking plug is inserted. The thread in an aluminium alloy cylinder head is damaged very easily if the sparking plug is overtightened. If necessary, the thread can be reclaimed by fitting what is known as a Helicoil insert. Most agents have facilities for this type of repair, which is not expensive.

4 If the cylinder head joint has shown signs of oil seepage when the machine was in use, check whether the cylinder head is distorted by laying it on a sheet of plate glass. Severe distortion will necessitate a replacement head but if the distortion is only slight it is permissible to wrap some emery cloth (fine grade) around the sheet of glass and rub down the joint using a rotary motion, until it is once again flat. The usual cause of distortion is uneven tightening of the cylinder head nuts.

5 The early D1 and D3 Competition models have what is known as a decompressor fitted to the cylinder head, which is operated from a trigger lever on the handlebars. The decompressor takes the form of a spring loaded poppet valve that vents the compressed petrol/air mixture to the atmosphere when the valve is depressed. The complete unit screws into the cylinder head in the same manner as the sparking plug.

6 Check to ensure the decompressor valve is not leaking and that the cable adjustment is correct. If the decompressor valve and seat are pitted or marked, they can be ground together after dismantling using fine valve grinding paste. It is unlikely that any of the parts will need replacement since the decompressor is normally used only when starting or stopping the engine, or when descending very steep inclines such as those encountered in the average trials event.

23 Crankcases - Examination and Renovation

1 Inspect the crankcases for cracks or any other signs of damage. If a crack is found, specialist treatment will be required to effect a satisfactory repair.

2 Clean off the jointing faces, using a rag soaked in methylated spirit to remove old gasket cement. Do not use a scraper because the jointing surfaces are damaged very easily. A leak-tight crankcase is an essential requirement of any two-stroke engine. Check also the bearing housings, to make sure they are not damaged. The entry to the housings should be free from burrs or lips.

3 Do not forget to check also the generator inner and outer covers and the primary drive cover. Good jointing surfaces are essential especially in the case of the generator inner and outer covers that have no intermediate gasket.

24 Gearbox Components - Examination and Renovation

1 Examine carefully the gearbox components for signs of wear or damage such as chipped or broken teeth on the gear pinions and kickstarter quadrant, rounded dogs on the ends of the gear pinions, bent selector forks, weakened or damaged springs and worn splines. If there is any doubt about the condition of a part, it is preferable to play safe and replace the part at this stage. Remember that if a suspect part should fail later, it will be necessary to completely strip the engine/gear unit yet again.

2 It is advisable to replace the kickstarter return spring irrespective of whether it seems to be in good condition. This spring is in constant use, yet if it has to be replaced at a later date, a certain amount of dismantling is necessary in order to gain access. It is cheap and easy to replace at this stage.

3 Do not forget to examine also the kickstarter ratchet assembly, which is held to the back of the clutch drum assembly by a circlip. Examination will show whether the ratchet teeth have worn, causing the kickstarter to slip or whether the outer teeth are damaged, causing the kickstarter quadrant to jam. Note that the leading tooth of the quadrant is relieved, to help offset the tendency to jam during the initial engagement.

25 Clutch Actuating Mechanism - Examination

1 The clutch actuating mechanism and adjuster are attached to the rear of the generator inner cover. It is unlikely that these parts will require attention, particularly if the actuating mechanism has been greased regularly during routine maintenance.

2 Should it be necessary to dismantle the mechanism, unclip

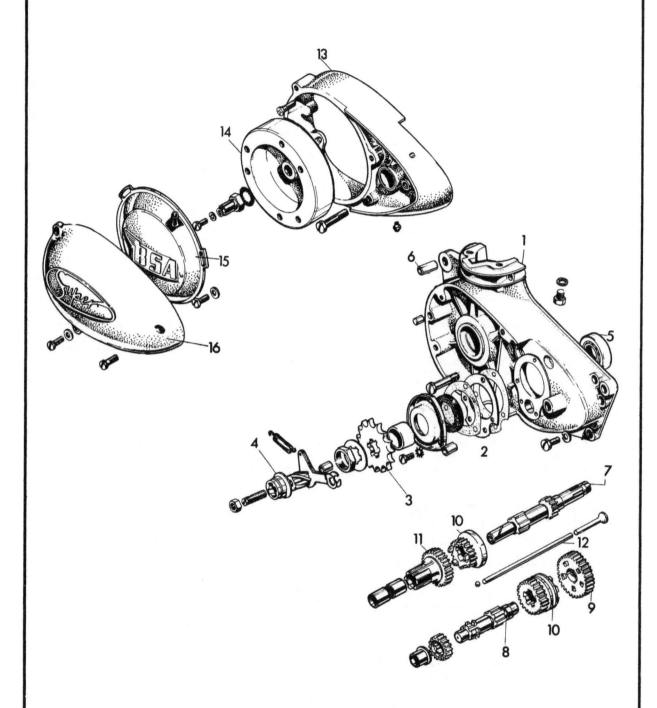

FIG. 1.6. LEFT HAND CRANKCASE AND GEARBOX COMPONENTS (D7)

1 Left hand crankcase	5 Gearbox main bearing	10 Second gear pinions (sliding)	13 Left hand crankcase cover
2 Final drive oil seal assembly	6 Crankcase dowel	11 Top gear sleeve pinion	14 Flywheel generator rotor
3 Final drive sprocket	7 Gearbox mainshaft	12 Clutch push rod assembly	15 Flywheel generator cover
4 Clutch operating mechanism	8 Gearbox layshaft		16 Left hand crankcase outer cover
	9 Bottom gear pinion		

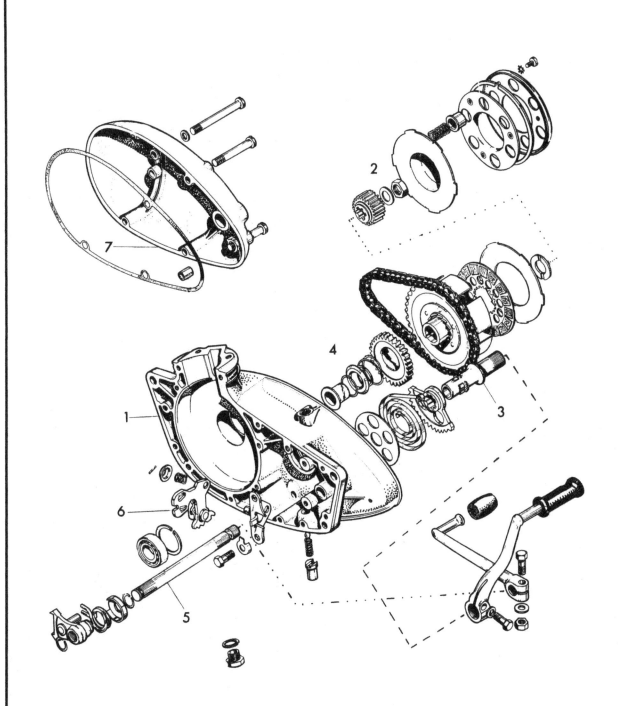

FIG. 1.7. RIGHT HAND CRANKCASE AND CLUTCH COMPONENTS (D7)

1 Right hand crankcase	3 Kickstarter shaft	5 Gear change lever shaft	7 Right hand outer cover
2 Clutch assembly	4 Kickstarter ratchet assembly	6 Gear indexing mechanism	

Fig.1.8. Clutch component parts

the return spring, unscrew the adjuster locknut and press the actuating lever out of its bush. The lever works on the quick-start worm principle and is unlikely to give trouble unless it is under-lubricated. The clutch push rod ball and rubber sleeve are inserted loosely in the end of the lever boss.

26 Clutch Assembly - Examination and Renovation

1 Examine the condition of the linings of the inserted clutch plates. If they are damaged, loose or have worn thin, replacements will be required.
2 Examine the tongues of the plain clutch plates, where they engage with the clutch drum. After an extended period of service, burrs will form on the edges of the tongues which will correspond with grooves worn in the clutch drum slots. These burrs must be removed, by dressing with a smooth file.
3 The grooves worn in the clutch drum slots can be dressed in a similar manner, making sure that the edges of the slots are square once again. If this simple operation is overlooked, clutch troubles will persist because the plates tend to lodge in the grooves when the clutch is withdrawn and promote clutch drag.
4 Check also the condition of the clutch springs. They should have a free length of 1.2025 inch and must be replaced if they compress much below this figure.
5 The clutch push rod is in two sections, a long section of approximately 5¾ inches and a short length of approximately 2 inches. Replace either section of the rod if the ends show a tendency to bell out. This is usually a sign of insufficient free play in the actuating mechanism, which has caused the ends to press together and generate heat, which in turn has destroyed the temper and caused them to soften and wear rapidly. The need for continuous clutch adjustment can invariably be traced to this type of fault.

27 Primary Chain - Examination and Replacement

1 Although the primary chain runs in ideal conditions, where it is both enclosed and fully lubricated, inspection is necessary from time to time. No means of adjustment are provided; when the chain becomes too slack it has reached the point of renewal and must be replaced.
2 Always fit the spring link so that the closed end faces the direction of travel. It is easier to fit the spring link assembly if the chain is pressed into one of the sprockets at the point of attachment. Never replace a damaged or bent spring clip.
3 After a considerable period of service, it is probable that the sprockets will also need renewing. The usual indication occurs when a new primary chain is fitted which will tend to be too slack as the result of sprocket wear. When the sprockets have to be replaced, renew also the chain. It is bad practice to use an old chain with new sprockets - all should be renewed and run together at the same time.
4 The primary chain tension is correct when there is approximately 3/8 inch play in the middle of the run. The chain should be replaced when the amount of play exceeds ¾ inch.

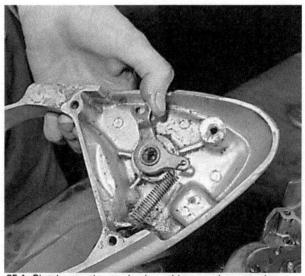

25.1. Clutch actuating mechanism seldom requires attention

28 Reassembly - General

1 Before the engine, clutch and gearbox components are re-assembled, they must be cleaned thoroughly so that all traces of old oil, sludge, dirt and gaskets are removed. Wipe each part clean with a dry, lint-free rag to make sure that there is nothing to block the internal oilways of the engine.

2 Lay out all the spanners and other tools likely to be required so that they are close at hand during the reassembly sequence. Make sure the new gaskets and oil seals are available - there is nothing more infuriating then having to stop in the middle of a reassembly sequence because a gasket or some other vital component has been overlooked.

3 Make sure the reassembly area is clean and unobstructed and that an oil can with clean engine oil is available so that the parts can be lubricated before they are reassembled. Refer back to the torque wrench settings and clearance data where necessary. Never guess or take a chance when this data is available.

4 Do not rush the reassembly operation or follow the instructions out of sequence. Above all, do not use excess force when parts will not fit together correctly. There is invariably good reason why they will not fit, often because the wrong method of assembly has been used.

29 Engine Reassembly - Fitting Bearings to Crankcase

1 Before fitting the crankcase bearings, make sure that the bearing surfaces are scrupulously clean and that there are no burrs or lips on the entry to the housings. Press or drive the bearings into the cases, using a mandrel and hammer, after first making sure that they are lined up squarely. Warming the crankcases will help when a bearing is a particularly tight fit.

2 When the bearings have been driven home, lightly oil them and make sure they revolve freely. This is particularly important in the case of the main bearings. There are two bearings on the drive side and one bearing on the generator side (right hand and left hand crankcases respectively) each of the journal ball type.

3 Using a soft mandrel, drive the oil seals into their respective locations. Do not use more force than is necessary because the seals will damage very easily. Good crankcase seals are essential to the efficient running of any two-stroke engine and if there is any doubt about the condition of the old seals they should be replaced without hesitation. Poor starting and indifferent running can often be attributed to worn or damaged oil seals, that allow air to enter the crankcase and dilute the incoming mixture whilst it is under crankcase compression.

4 Do not omit the circlips that retain the bearings in position.

30 Engine Reassembly - Right Hand Crankcase

1 Place the right hand crankcase on two wooden blocks or in an engine stand so that the inner side faces upwards.

2 Assemble the gear clusters on their respective shafts, following the order shown in the accompanying illustration, (three speed models only). If it has been removed, fit the cam plate plunger housing, spring and ball bearing. Replace the gear quadrant selector making sure the selector engages with the grooved track on the layshaft second gear pinion. Check also that the cam plate ball bearing locates with one of the notches in the quadrant. Tighten the two retaining bolts and bend over the tab washers to lock them in position.

3 Oil the gear change lever spindle and slide it into its housing. Push it home fully with the dowel pin centrally disposed in the slot of the gear quadrant selector mounting plate. The claw should locate around the selector itself.

4 Before inserting the gear change lever spindle and claw assembly, check that the gear lever return spring is in good condition. In many respects it is wise to replace the spring irrespective of its appearance, especially in view of its low cost. A complete engine strip is necessary in the event of a spring breakage. The spring is located immediately behind the selector claw.

30.2A. Three-speed gear cluster in assembled form

30.2B. Fit assembly as complete unit

30.2C. Replace cam plate plunger ball

5   Before proceeding with the engine reassembly, check that the gear change mechanism functions correctly and that each of the gears engage fully. It will be necessary to turn the shafts to permit correct engagement of the sliding dogs.

6   Due to certain design changes, necessary when the four-speed gearbox was introduced, a somewhat different gearbox reassembly procedure is recommended for the D10 and D14 models only. The following revised instructions should be adopted:

7   Refit the cam plate plunger housing, spring and ball. Place the layshaft first gear in position. Assemble the gear selector forks, using grease to retain the rollers. The forks are 'handed' and when the two flat edges are placed together, the rollers should line up towards the top.

8   Assemble the cam plate with the mainshaft sliding gear in the uppermost selector fork (second gear pinion). Insert the mainshaft complete with first and third gears and slide the selector assembly into position so that the sliding gear engages with the splines on the mainshaft. The sliding gear should be positioned so that the track that engages with the uppermost selector fork faces downwards. Do not omit the circlip that holds the mainshaft third gear in position, before the sliding gear is inserted.

9   Replace the cam plate mounting bracket and secure it with the two bolts and tab washers. Check that the rollers are engaged with the cam tracks and function correctly.

10 Fit the layshaft sliding gear in the lower selector fork, facing the reverse direction to the sliding gear on the mainshaft. The layshaft can now be inserted, complete with top gear and second gear, the latter retained by a circlip.

11 Fit the gearchange lever return spring to the gearchange spindle, locating the ends over a projection on the cam plate mounting bracket. Place the distance piece below the spring and replace the spindle in its housing after oiling the stem.

12 Replace the top gear pinion on the mainshaft. If desired, this sleeve pinion can be fitted to the left hand crankcase and the oil seal assembly and final drive sprocket attached, prior to the joining of the crankcases. See Section 35 of this Chapter for the assembly details.

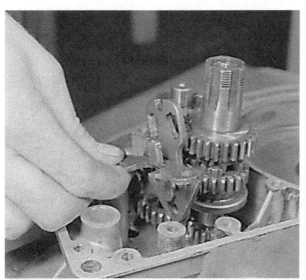

30.2D. Engage camplate quadrant with plunger ball first

30.2E. Tab washers secure bolts

30.3A. Oil gear lever spindle prior to insertion

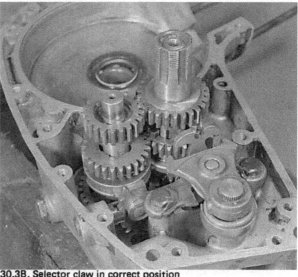

30.3B. Selector claw in correct position

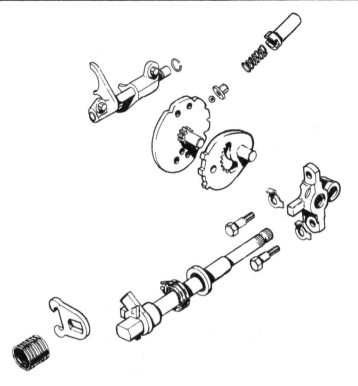

Fig.1.9. Gear change mechanism components

31 Engine Reassembly - Fitting the Crankshaft Assembly

1 Before fitting the crankshaft assembly to the right hand crankcase, check that there are shims of equal thickness on both sides of the crankshaft. These shims take up the end float of the crankshaft assembly and must be evenly distributed so that the connecting rod is centrally disposed within the crankcases.

2 Apply a liberal coating of oil to the crankshaft and refit the assembly to the right hand crankcase.

3 Lightly coat both crankcases with gasket cement after making sure that the mating surfaces are clean and undamaged. A good air-tight joint is ESSENTIAL to the correct running of the engine. Lower the left hand crankcase on to the right hand built-up assembly and replace the 16 cross-head screws that clamp both crankcases together (11 only, early models). Note that one screw is fitted from within the right hand crankcase, above and to the right of the gearbox main bearing.

4 Great care should be exercised when refitting the crankshaft assembly in order to avoid damaging the oil seals. Grease the inside of the oil seals and check that the crankshafts have a liberal coating of oil before fitting. Avoid the use of force at all costs.

5 Check that the end float of the crankshaft assembly is within the limits of 0.004 in. to 0.006 in. Shims must be added or removed from the crankshaft assembly until the correct amount of end float is achieved. There must be an equal amount of shims on BOTH sides of the flywheel assembly.

32 Engine Reassembly - Fitting the Kickstarter Assembly

1 Invert the crankcase assembly so that the right hand side faces upwards. Fit the large perforated washer, clock-type return spring and kickstarter shaft complete with quadrant.

2 Loop the end of the return spring over the projection within the crankcase and tension the spring by rotating the kickstarter shaft one turn in the anti-clockwise direction. The quadrant should rest against the abutment towards the rear of the crankcase, which acts as a stop.

33 Engine Reassembly - Fitting the Clutch

1 Before fitting the clutch to the gearbox mainshaft, oil and replace the phosphor bronze bush that slides over the mainshaft splines. Fit the clutch drum, to which the kickstarter pinion and ratchet assembly are attached at the rear. These latter parts seldom require attention; if wear causes the ratchet to slip or the quadrant to jam, replacement of all the defective parts is necessary.

2 Replace the phosphor bronze thrust washer that locates over the mainshaft splines and slide the clutch centre down the splines until it rests on this washer. Replace the washer and retaining nut and tighten the latter fully. Grease and insert the clutch operating pushrod with the mushroom head, into the hollow mainshaft.

3 Replace the clutch plates, commencing with an inserted plate. Alternate with plain and inserted plates and end with the pressure plate that has a domed centre.

4 Fit the end plate complete with the six clutch springs and the cups into which they fit. Compress this plate using the second pair of hands of an assistant or by adapting a sprocket puller in the method illustrated. Whilst the plate is under tension, insert the retaining circlip, checking to ensure that it engages fully with the groove cut within the clutch drum. Release the pressure and replace the end cap that is held in position by three screws. The clutch is now complete.

34 Engine Reassembly - Fitting the Engine Sprocket and Primary Chain

1 The engine sprocket is keyed on to a taper on the right hand crankshaft. Replace the woodruff key and lower the sprocket into position. It is retained by a nut and tab washer. Before tightening the nut, fit the primary chain. Reconnection is made easy if the chain is held in the teeth of the clutch sprocket whilst the spring link is fitted. Make sure that the spring link locates with the grooves in the link assembly and that the closed end is facing in the direction of travel when the engine is running.

2 Lock the clutch and/or engine sprocket and tighten the sprocket retaining nut.

31.1. Shims limit endfloat to correct tolerances

31.2. Oil crankshafts before refitting flywheel assembly

31.3A. Only a light coating of gasket cement is necessary

31.3B. One retaining screw is in right hand crankcase

32.1. Kickstarter assembly prior to tensioning spring

33.1. Clutch drum has ratchet assembly on back

33.2. Do not forget clutch operating 'mushroom'

33.3A. Correct assembly order of clutch plates

33.3B. Pressure plate fits next ...

33.4A. Then end plate with springs and cups

33.4B. Insert retaining circlip whilst clutch is compressed

34.1A. Engine sprocket is keyed on crankshaft

3   The D10 and D14 models employ a self-locking nut; there is no tab washer. After this nut has been tightened fully, the contact breaker cam can be loosely assembled on the end of the crankshaft.

4   Using a new gasket, refit the right hand crankcase cover and tighten the five retaining screws.

35 Engine Reassembly - Fitting the Final Drive Sprocket and Oil Seal

1   Re-invert the crankcase assembly so that the left hand side faces upwards. Place a new gasket over the flange surrounding the gearbox main bearing and place the bearing retaining plate above it. Add also the plate that contains the oil seal for the projecting gear pinion sleeve. Lubricate the sleeve and grease also the inner portion of the oil seal so that it is not damaged when the oil seal retaining plate is lowered into position.

2   Before the retaining screws (five) are tightened fully, slide the distance piece over the gear pinion sleeve and replace the final drive sprocket. Tighten the screws, then add the tab washer and nut that secure the final drive sprocket to the gear pinion sleeve. This nut has a LEFT HAND thread.

3   Lock the sprocket with a length of chain held in a vice or by wrapping the chain around the crankcase assembly, and tighten the nut to a torque setting of 50 - 55 ft lbs. Bend the tab washer so that it retains the nut in position.

36 Engine Reassembly - Fitting the Flywheel Magneto Generator

1   Rest the engine unit on its side with the left hand crankcase facing upwards. Refit the left hand crankcase outer cover after greasing and replacing the clutch push rod in the hollow mainshaft. If the rod is bent or the hardened ends have worn soft, fit a new replacement.

2   There is no gasket fitted between the left hand crankcase and the left hand cover. The joint should be made dry. Replace all the retaining screws, some of which remain captive when the flywheel rotor is fitted.

3   Check clutch action before proceeding further by turning the adjusting nut clockwise with a spanner. Spring resistance should be felt.

4   Replace the woodruff key in the crankshaft taper and lower the flywheel rotor into position. Make sure it seats firmly on the taper before fitting the retaining nut and washer. Tighten to a torque setting of 50 - 55 ft lbs and avoid using excessive force or sudden blows. The long mainshaft is bent very easily unless care is taken.

5   The D10 and D14 models have a somewhat similar arrangement, although the mainshaft is keyed but not tapered.

6   A convenient method of locking the engine whilst the rotor nut is tightened is to place a metal rod through the small end of the connecting rod so that it rests across the crankcase mouth.

7   Replace the stator plate assembly after oiling the centre bush. On the D1, D3, D5 and D7 models it is retained by three screws that pass through slots around the edge of the plate. Only one of these screws - the one nearest to the final drive sprocket - can be replaced before the domed end cover is fitted. The very early D1 models and those fitted with the early Lucas ignition equipment differ in minor respects only.

8   Before the stator plate can be fitted to the D10 and D14 models, distance pieces must be fitted to the three studs that project from the left hand crankcase cover. The stator cable should be in the 'three o'clock' position if the stator is positioned correctly. Replace and tighten the retaining nuts and their spring washers, with the extra-long nut furthest forward. Use a feeler gauge to check that the rotor is an equal distance from each of the stator coils; if the distance varies, slacken off the retaining nuts and reposition the stator assembly until the discrepencies disappear.

9   Locate the cable grommet into its recess in the inner cover. On the earlier models, the cable passes through a hole in the top

34.1B. Check tapers are undamaged before refitting sprocket

34.1C. Reconnect chain at clutch sprocket

34.2. Nut should be tightened with engine locked

35.1A. Fit new gasket to flange ...

35.1B.   ... then retaining plate ...

35.1C. ...followed by plate containing oil seal

35.2A. Add distance piece before tightening screws

35.2B. Sprocket fits on splines

35.2C. Securing nut has a LEFT HAND thread. Do not omit tab washer

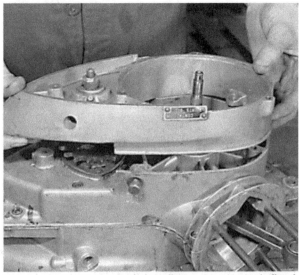

36.1. Outer cover must be replaced before generator rotor is fitted

36.4A. Generator rotor is keyed onto crankshaft

36.4B. Check tapers are undamaged before fitting

36.4C. Plain washer fits behind retaining nut

36.7. Centre bush of stator plate is easily overlooked. Oil before fitting

37.2. Warm piston if gudgeon pin is a very tight fit

of the left hand crankcase cover.

10 Replace the domed outer cover which is retained by four cross head screws. A blanking-off cap is a push fit into the cover, to provide access for clutch adjustment (D10 and D14 models only).

37 Engine Reassembly - Refitting the Piston and Cylinder Barrel

1 Refit the piston complete with piston rings to the connecting rod. If the advice given for dismantling the engine has been followed, the inside of the piston skirt will have been marked so that the piston is fitted in the same position. If the piston was not marked, it should be refitted so that the piston ring gaps are closest to the FRONT of the engine.

2 If the gudgeon pin is a tight fit, it may be necessary to warm the piston in order to expand the gudgeon pin bosses. Holding the piston under a stream of hot water will usually suffice, provided the water is quickly wiped off before reassembly commences. Push the pin home and replace both circlips, which should have been renewed. It is false economy to reuse the originals, which may become displaced and cause serious engine damage.

3 It is wise to pad the mouth of the crankcase with rag whilst the circlips are being fitted, to prevent them from falling in. If a misplaced circlip drops in, it may be necessary to dismantle the complete engine in order to reclaim it.

4 Check that both circlips are located in their grooves.

5 Before the cylinder barrel is replaced, check that the two screws at the mouth of the crankcase spigot are loose. Oil the piston and the inside of the cylinder barrel. Note that if the engine has been rebored it will probably be necessary to relieve the edges of the ports very slightly to prevent the rings from being trapped. Careful work with emery cloth and/or a fine file will remove any burrs or lips; make sure none of the abrasive material remains in the cylinder bore before it is wiped clean and coated with oil. Fit a new cylinder base gasket.

6 The piston will have two or three rings, depending on the model and general specification. Note that the ring grooves are fitted with a peg, which predetermines where the ring gaps will occur. This is essential, otherwise the ends of the rings will project into the ports, causing breakages and severe engine damage.

7 Slide the cylinder barrel down the holding studs and feed the piston in by compressing one ring at a time. Check that the ends of each ring coincide with the pegs. The bottom of the cylinder barrel has a slight chamfer to aid the insertion of the rings.

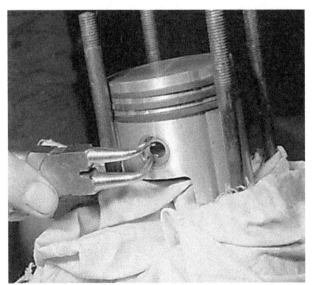

37.4. Circlips must seat correctly in grooves

8 It is permissible to use a piston ring clamp if this method is preferred, assuming care is taken to ensure each ring is located correctly with its peg. Lower the cylinder barrel until it seats on the new base gasket.

9 Fit a new cylinder head gasket and lower the cylinder head into position. The cylinder head nuts should be tightened down in a diagonal sequence, a little at a time, until a final torque setting of 18 - 22 ft lbs is achieved. Complete the assembly by tightening the two cross head screws at the base of the cylinder barrel spigot.

10 Before proceeding to the next stage of reassembly, check that the engine turns freely in either direction. Any tight spots must be investigated and remedied, even if it means removing the cylinder barrel and head again.

38 Engine Reassembly - Fitting the Contact Breaker Assembly

D10 and D14 models only:

1 On these models, the contact breaker assembly is fitted within a cast-in housing of the right hand crankcase cover. Replace the contact breaker plate assembly by replacing the two cross head screws in two of the opposing slots (5 o'clock and 11 o'clock positions respectively). Make sure the contact breaker cam is home on the centre shaft, without tightening the centre retaining screw at this stage. Rotate the engine until the contact breaker points are open fully and check that the gap is 0.012 in. If necessary, adjust the points until this gap is achieved, by following the procedure in Chapter 3, section 4.

2 Time the ignition at 16½° before top dead centre, (BTDC), using the timing disc procedure described fully in Chapter 3, section 7.

3 After a double check that the timing is correct, replace the domed left hand crankcase outer cover after detaching the timing disc and re-tightening the rotor retaining nut. Replace also the circular cover that seals off the contact breaker assembly in the right hand crankcase cover.

D1, D3, D5 and D7 models

1 On these models the contact breaker assembly is located within the flywheel magneto stator plate housing. With the exception of the very early models fitted with Lucas ignition, the contact breaker plate assembly is integral with the stator plate casting.

2 Attach the contact breaker cam to the end of the crankshaft. Its position is predetermined by a keyway. Tightening the centre screw secures the cam on the shaft and also forces the key into the mating keyway.

3 The cam on the early Lucas ignition models press fits on to the end of the crankshaft. The location is critical, particularly with regard to the correct functioning of the "Emergency Start" procedure. Marks on both the cam and the shaft ensure correct alignment. Internal threads in the cam itself can be used as a means of extraction if the setting is incorrect.

4 Check that the contact breaker points are gapped at 0.012 in when they are open fully. If necessary, adjust by following the procedure detailed in Chapter 3, section 4.

5 Time the ignition at 5/16 in before top dead centre (BTDC) (D1 and D3 models) or 1/16 in before top dead centre (D5 and D7 models) using the procedure detailed in Chapter 3, section 7.

6 After a double check that the ignition timing is correct, oil the wick that bears on contact breaker cam and replace the circular cover that seals off the contact breaker assembly. (Circular cover plate and domed outer cover, D5 and D7 models).

39 Refitting the Engine/Gearbox Unit in the Frame

1 Follow in reverse the procedure given in section 5 of this Chapter. Fit the sparking plug as soon as the engine unit is lifted into the frame and the engine bolts are replaced, to prevent dirt and foreign matter from dropping into the engine.

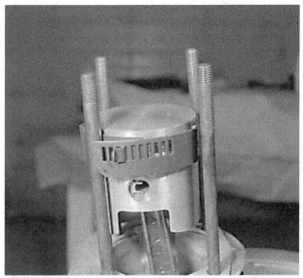

37.8A. Ring clamp can be used when rings are correctly aligned

37.8B. Check rings are not trapped when lowering cylinder barrel

37.9A. Never re-use the old cylinder head gasket. Always fit a new replacement

37.9B. Cylinder head flange must be clean and flat

37.9C. Finish by tightening screws at crankcase mouth

38.2A. Cam fits on end of crankshaft

38.2B. Retaining screw forces key into keyway of cam

38.6A. Oil wick of contact breaker cam

38.6B. Circular cover seals off contact breaker assembly

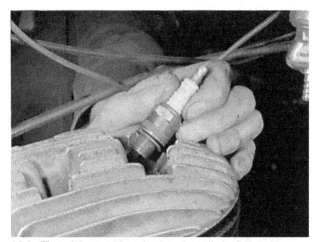

39.1. Thread the sparking plug into the cylinder head - it's a good idea to fit a new plug

2 Use a new copper/asbestos ring gasket for the exhaust pipe joint. A leaktight exhaust system is essential for the correct running of the engine.

40 Completion of Reassembly and Final Adjustments

1 Before attaching the carburettor, make sure the 'O' ring seal in the centre of the flange is in good condition and is located correctly ('Monobloc' and 'Concentric' carburettors only). Check also that the nipple at the end of the throttle cable seats correctly within the throttle slide, otherwise the throttle may stick open and cause the engine to race.

2 Refit the kickstarter and gear change levers, so that they are positioned at the correct angles. These are easier to determine when the engine is in the frame.

3 Refit the clutch cable by operating the clutch worm with a spanner. This will ensure there is sufficient slack in the cable to slip the cable end into the cable stop, under the left hand crankcase.

4 Remove the gearbox oil filler cap/dipstick from the top of the right-hand crankcase and refill the gearbox with SAE 40 motor oil. On 3-speed models, the filler cap has an integral dipstick – the oil level is correct when it reaches the base of the dipstick with the cap/dipstick just resting on the crankcase hole, or when it reaches the groove on the dipstick with the cap/dipstick fully in place in the crankcase. On 4-speed models, remove the level screw set in the side of the right-hand crankcase (see page 8) and add oil until it begins to flow out of the level screw hole, then refit the screw. In all cases, check the level with the motorcycle upright on level ground and do not overfill.

41 Starting and Running the Rebuilt Engine

1 When the initial start-up is made, run the engine slowly for the first few minutes, especially if the engine has been rebored. Check that all the controls function correctly and that there are no oil leaks before taking the machine on the road.

2 Remember that a good seal between the piston and the cylinder barrel is essential for the correct functioning of any two-stroke engine. In consequence, a rebored engine will require more careful running-in than its four-stroke counterpart. There is a far greater risk of engine seizure during the first hundred miles if the engine is permitted to work hard.

3 Do not add extra oil to the petrol/oil mix in the mistaken belief that it will aid running in. More oil means less petrol and the engine will run with a permanently weakened mixture, causing overheating and a far greater risk of engine seizure. Keep to the recommended proportions.

4 Do not tamper with the exhaust system or run the engine without the baffles fitted to the silencer. Unwarranted changes in the exhaust system will have a very noticeable effect on engine performance, invariably for the worst.

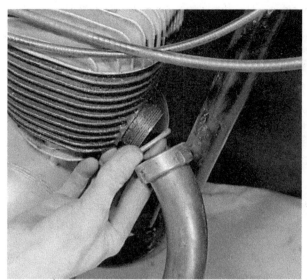

39.2. New exhaust pipe gasket is essential

40.1A. 'O' ring seal must be in good shape and located correctly

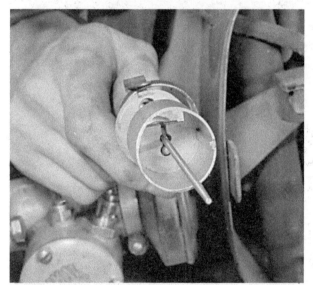

40.1B, Make sure nipple seats correctly in slide

40.2A. Kickstarter arm fits first ...

40.2B. ...then gear change lever

40.3. Operate clutch worm to gain slack in cable

40.4A. Fill gearbox after removing cap

40.4B. Cap has integral dipstick

42 Fault Diagnosis - Engine

Symptom	Cause	Remedy
Engine will not start	Defective sparking plug	Remove plug and lay on cylinder head. Check whether spark occurs when engine is kicked over.
	Dirty or closed contact breaker points	Check condition of points and whether gap is correct.
	Air leak at crankcase or worn oil seals around crankshaft	Flood carburettor and check whether mixture is reaching the sparking plug.
	Clutch slip	Check and adjust clutch.
Engine runs unevenly	Ignition and/or fuel system fault	Check systems as though engine will not start.
	Blowing cylinder head gasket	Leak should be evident from oil leakage where gas escapes.
	Incorrect ignition timing	Check timing and reset if necessary.
	Loose pin on which moving contact breaker point pivots	Replace defective parts.
Lack of power	Incorrect ignition timing	See above.
	Fault in fuel system	Check system and filler cap vent.
	Blowing head gasket	See above.
	Choked silencer	Clean out baffles.
High fuel/oil consumption	Cylinder barrel in need of rebore and o/s piston	Fit new rings and piston after rebore.
	Oil leaks or air leaks from damaged gaskets or oil seals	Trace source of leak and replace damaged gaskets or seals.
Excessive mechanical noise	Worn cylinder barrel (piston slap)	Rebore and fit o/s piston.
	Worn small end bearing (rattle)	Replace bearing and gudgeon pin.
	Worn big-end bearing (knock)	Fit new big-end bearing.
	Worn main bearings (rumble)	Fit new journal bearings and seals.
Engine overheats and fades	Pre-ignition and/or weak mixture	Check carburettor settings. Check also whether plug grade correct.
	Lubrication failure	Is correct measure of oil mixed with petrol?

43 Fault Diagnosis - Clutch

Symptom	Cause	Remedy
Engine speed increases but machine does not respond	Clutch slip	Check clutch adjustment for pressure on pushrod. Also free play at handlebar lever. Check condition of clutch plate linings, also free length of clutch springs. Replace if necessary.
Difficulty in engaging gears. Gear changes jerky and machine creeps forward, even when clutch is fully withdrawn	Clutch drag	Check clutch adjustment for too much free play.
	Clutch plates worn and/or clutch drum	Check for burrs on clutch plate tongues or indentations in clutch drum slots. Dress with file.
	Clutch assembly loose on mainshaft	Check tightness of retaining nut. If loose, fit new tab washer and retighten.
Operating action stiff	Damaged, trapped or frayed control cable	Check cable and replace if necessary. Make sure cable is lubricated and has no sharp bends.
	Bent pushrod	Replace.

44 Fault Diagnosis - Gearbox

Symptom	Cause	Remedy
Difficulty in engaging gears	Gear selectors not indexed correctly	Check alignment of 'timing' marks on selector assembly.
	Gear selector forks bent	Replace.
	Broken or misplaced selector springs	Replace broken springs and re-locate as necessary.
Machine jumps out of gear	Worn dogs on ends of gear pinions	Replace worn pinions.
	Cam plate plunger stuck	Free plunger assembly.
Kickstarter does not return when engine is turned over or started	Broken or badly tensioned kickstarter return spring	Replace spring or retension.
Gear change lever does not return to normal position	Broken return spring	Replace.

Chapter 2 Fuel system and carburation

Contents

Specifications

Petrol Tank Capacity

D1, D3, D5 and D7 models	1 3/4 gallons (8 litres)
D10 and D14 models	1 7/8 gallons (8.3 litres)

Carburettor

Make		Amal - all models	
Type		261/001D	(D1 models 1948 - 50)
		361/1	(D1 models 1951 - 59)
		223/7	(D3 models)
	'Monobloc'	375/31	(D5 and D7 models)
	'Monobloc'	376/323	(D10 models)
	'Concentric'	R626/2	(D10 and D14 models)
	'Concentric'	R626/17	('Bushman' models)
Main jet		75	(D1 models)
		95	(D3 models)
		140	(D5 and D7 models)
		180 (376/323)	(All other models)
		150 (R626/2)	(All other models)
		180 (R626/17)	(All other models)
Pilot jet		Fixed	(D1 and D3 models)
		25	(D5 and D7 models)
		622/107	(D10 and D14 models)
Throttle slide		5	(D1 and D3 models)
		3 or 3 1/2	(All other models)
Needle jet		0.106	(D1 models)
		0.1075	(D3 models)
		0.105	(All other models)
Needle position		2	(All models)

Note that the carburettor settings for the 'Bushman' model are identical to those recommended for the D14 models, despite the different carburettor specification.

1 General Description

The fuel system comprises a petrol tank from which a petrol/oil mix of controlled proportions is fed by gravity to the float chamber of the Amal carburettor. A petrol tap with a built-in gauze filter is located beneath the rear end of the petrol tank, which on all except D1 models has provision for turning on a small reserve quantity of fuel when the main content of the tank is exhausted. There is an additional filter within the main feed union of the 'Monobloc' and 'Concentric' carburettors, fitted to the later models only.

For cold starting purposes, the carburettor fitted to the D1, D3, D5 and D7 models incorporates a hand-operated strangler attached to the air intake. A series of vanes closes off the air intake, to give the rich mixture needed for a cold start. As soon as the engine has started, the strangler can be opened until the engine will accept full air as under normal running conditions. On later models, fitted with a 'Concentric' carburettor, a handlebar control operates a moveable air slide that performs the same function.

2 Petrol/Oil mix - Correct Ratio

1 The engine uses the 'petroil' system of lubrication, requiring a measured amount of self-mixing two-stroke oil to be added to each gallon of petrol. The original mixing ratios specified by BSA

at the time of manufacture are given below and based on the two-stroke oils available at the time – modern two-stroke oils may require a revised mixture. The ratio for D1, D3, D5 and D7 models is 20:1 (twenty parts petrol to one part oil). Later models (D10, D14/4 and D175) use a 24:1 petrol/oil ratio.

2 The fuel tank filler cap incorporates an oil measuring cup, which enables the above ratios to be obtained when mixed as follows:

D1, D3, D5 and D7 models
    with 5 inch long filler cap   2½ cups oil to 1 gallon petrol
    with 6? inch long filler cap   2 cups oil to 1 gallon petrol
D10, D14/4 and D175   3½ cups oil to 1 gallon petrol

3 It is recommended that the oil and fuel mixture is prepared in an auxiliary fuel can to ensure the correct ratio is obtained. Shake the can well before pouring the fuel mix into the motorcycle's tank.

4 Engine lubricant is dependent solely on the oil content of the fuel mixture which passes through the carburettor and into the engine. Never add fresh fuel without a measure of oil otherwise the engine will seize due to lack of lubrication. Avoid prolonged coasting downhill with the throttle closed as engine lubrication will be cut-off.

3 Petrol Tank - Removal and Replacement

1 It is unlikely that there will be need to remove the petrol tank completely unless the machine has been laid up and rust has formed inside or it needs reconditioning. The engine/gearbox unit can be removed from the frame on all models without having to detach the tank completely; in the case of the D1, D3, D5 and D7 models there is sufficient clearance without need even to raise the tank.

2 The petrol tank is secured to the frame by two short bolts at the front, one on each side, that locate with tapped holes in the steering head casting. The rear mounting consists of a long bolt that passes through a lug attached to the top frame tube (below frame tube, early models) and through lugs welded to the end of the petrol tank. When all three bolts are removed, the tank can be lifted away without draining.

4 Petrol Tap - Removal and Replacement

1 The petrol tap is threaded into an insert in the left-hand side of the petrol tank. The early D1 models employ a push-pull tap that has no provision for reserve. Later models use a pull-out tap, the knob of which can be rotated anti-clockwise to bring in the small reserve quantity of fuel.

2 Before the tap can be unscrewed by applying a spanner to the flats close to the petrol tank joint, the tank must first be completely drained of fuel. When the tap is removed the gauze filter will be exposed, which is an integral part of the extension of the tap body.

3 On the later models only, the petrol feed pipe is a push fit on the tube at the base of the tap. Earlier models have a screwed union joint.

4 Replace by screwing the tap back into the tank, checking that the fibre sealing washer is in good condition.

5 Petrol Feed Pipe - Inspection

1 On the early models fitted with a push-pull petrol tap a

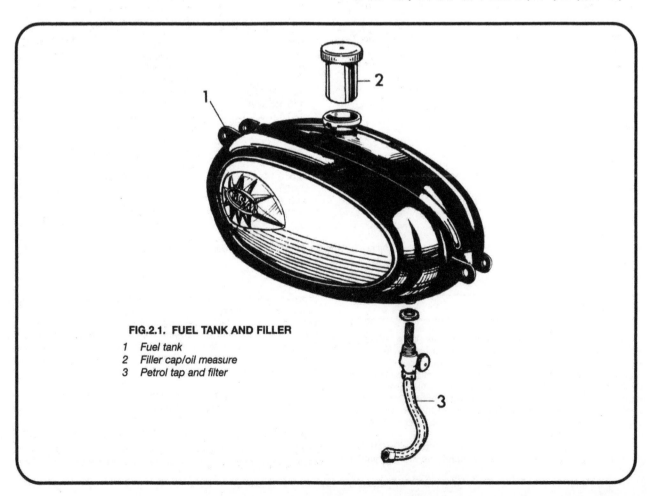

FIG.2.1. FUEL TANK AND FILLER

1   Fuel tank
2   Filler cap/oil measure
3   Petrol tap and filter

metal feed pipe was used with screwed union joints at each end. This arrangement is unlikely to give trouble unless a blockage occurs, part of the tube is kinked or if the unions come unsoldered. Replacement will be necessary in each case, unless the blockage can be cleared or the unions resoldered in position.

2 Later models use a plastic pipe that is a push fit on the end of the petrol tap outlet and has a union joint where connection is made with the carburettor float chamber. Replacement is necessary if the tube becomes hard or cracks, or if the push-on joint becomes slack.

6 Carburettor - Removal

1 Before removing the carburettor, detach the petrol feed pipe where it joins the float chamber, by unscrewing the union. Remove also the throttle slide and needle assembly by unscrewing the top of the mixing chamber and lifting it away complete with the control cable(s) and slide assembly. (Remove cross head retaining screws, "Concentric" carburettor only).

2 On the D1 and D3 models the carburettor is attached to the inlet stub of the cylinder barrel by a clip and pinch bolt arrangement. Slacken the pinch bolt and withdraw the carburettor complete. All other models have a flange joint, with two fixing bolts. If the bolts are withdrawn the carburettor will come free, complete with the 'O' ring seal in the centre of the carburettor flange.

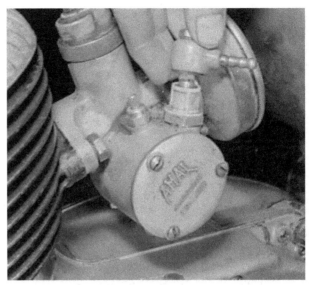

6.1A. Do not overlook the nylon filter within the feed pipe union

6.1B. Remove slide and needle assembly first

7 Carburettor - Dismantling and Inspection

1 To separate the float chamber unscrew and withdraw the large nut at the base of the carburettor mixing chamber (D1 and D3 models only). The float chamber is an integral part of the "Monobloc" carburettor fitted to the D5 and D7 models, and cannot be separated. In the case of the "Concentric" carburettor, fitted to the D10 and D14 models, the float chamber is held to the mixing chamber by two screws with spring washers, fitted from the underside.

2 Because the D1 and D3 models employed a top feed to the float chamber, it is necessary to remove also the float chamber top. Unscrew by applying a spanner across the flats at the base of the petrol feed union. Although the "Monobloc" float chamber cannot be detached, access to the internals is gained by removing the two slotted head screws in the side of the float chamber and detaching the cover and gasket.

3 Check the float needle and its seating for wear and whether the needle is bent. If there is a ridge around the needle and/or its seating, replacement will be needed.

4 Check the condition of the float and whether it has become porous so that petrol will leak inwards. The earlier copper floats

6.2. 'Monobloc' carburettor has a flange fitting

are more prone to this type of fault. A faulty float should be replaced, for it is not practicable to effect a permanent repair.

5 Check the throttle slide for wear and make sure none of the carburettor jets are blocked. Never use wire or any other pointed instrument to clear a blocked jet, otherwise there is risk of enlarging the jet and upsetting the carburation. Either blow the jet clear or use a blast of air from a tyre pump.

6 Make sure the float chamber body is clean and free from any sediment that may have originated from the petrol. Do not forget the tiny nylon gauze that is fitted within the petrol feed union of the "Monobloc" and "Concentric" carburettor only. This must be clean.

7 To reassemble the carburettor, follow the dismantling instructions in reverse. When replacing the float of the "Monobloc" carburettor on its hinge pin, do not fail to add the metal spacer. If this is omitted, the float can foul the end cover and stick, causing flooding and poor engine performance.

8 Before inserting the slide and needle assembly in the top of the mixing chamber, make quite sure the control cable nipples are correctly seated. If a cable end becomes misplaced the throttle slide will be held open making starting difficult and allowing the engine to race when it eventually runs.

9 Make sure the 'O' ring seal is in good shape and is positively

located, before making the flange joint with a "Monobloc" or a "Concentric" carburettor (D5, D7, D10 and D14 models). Gasket cement is not necessary at this joint.

10 Beware of overtightening the bolts on any of the flange-fitting carburettors because this will cause the flange to distort and bow, resulting in an air leak. If the flange is bowed already, remove the 'O' ring seal and rub down the flange with a sheet of emery cloth (fine grade) wrapped around a piece of glass. Apply the carburettor flange to this flat surface and rub with a rotary movement until the flange is again completely flat, as viewed against a straight edge. Make sure the carburettor is free from emery dust and replace the 'O' ring before refitting.

8 Carburettor - Checking the Settings

1   The various sizes of the jets and the throttle slide, needle and needle jet are predetermined by the manufacturer and should not require modification. Check with the Specifications list if there is any doubt about the values fitted.

2   Slow running is controlled by a combination of the throttle stop and pilot jet settings, irrespective of the type of carburettor fitted. Commence by screwing inwards the throttle stop screw so that the engine runs at a fast tick-over speed. Adjust the pilot jet screw until the tick-over is even, without either misfiring or hunting. Screw the throttle stop outwards again until the desired tick-over speed is obtained. Check again by turning the pilot jet screw until the tick-over is even. Always make these adjustments with the engine at normal working temperature and remember that the characteristics of a two-stroke engine are such that it is very difficult to secure an even tick-over at low engine speeds. Some prefer the engine to stop when the throttle is closed completely, but in a two-stroke engine with petroil lubrication there is always risk of oil starvation if the machine is coasting with the throttle closed.

3   As a rough guide, up to 1/8th throttle is controlled by the pilot jet, from 1/8th to ¼ throttle by the throttle slide cutaway, from ¼ to ¾ throttle by the needle position and from ¾ to full throttle by the size of the main jet. These are only approximate divisions; there is a certain amount of overlap.

4   The normal setting for the pilot jet screw is approximately one and a half full turns out from the fully closed position. If the engine 'dies' at low throttle openings, suspect a blocked pilot jet.

5   Guard against the possibility of incorrect carburettor adjustments that result in a weak mixture. Two-stroke engines are very susceptible to this type of fault, which will cause rapid overheating and subsequent engine seizure. Some owners believe that the addition of a little extra oil to the petrol will help prolong the life of the engine, whereas in practice quite the opposite occurs. Because there is more oil the petrol content is less and the engine runs with a permanently weakened mixture!

9 Air Cleaner - Location, Examination and Replacement of the Element

1   The D1, D3, D5 and D7 models have an air cleaner built into the air intake of the carburettor, behind the vanes of the hand-operated strangler. It is of the wire mesh type that requires dismantling and cleaning every 1,000 miles. Access is gained by unscrewing the pinch bolt through the clip that retains the intake unit in position and withdrawing the complete unit from the carburettor. Soak the complete unit in petrol, allow it to dry and then re-immerse it in engine oil for a few moments. Allow the surplus oil to drain off, wipe the exterior, and replace the unit on the carburettor.

2   The D10 and D14 models employ an air filter unit contained within the right-hand sidecover, below the nose of the dualseat. The lid of the cover is retained by two captive, chrome headed screws. When the lid is removed, the felt-covered wire mesh element will be exposed. It breathes through a wire mesh grille in the base of the cover and is connected to the carburettor air

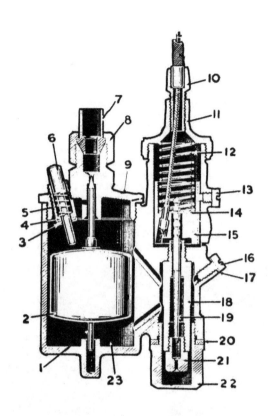

FIG.2.2. CROSS-SECTION THROUGH D1/D2 AMAL CARBURETTOR

1 Float needle	13 Throttle valve location
2 Float	screw
3 Tickler cotter	14 Jet needle clip
4 Tickler bush	15 Throttle valve
5 Tickler spring	16 Feed hole screw
6 Tickler	17 Feed hole washer
7 Petrol pipe union	18 Needle jet
8 Petrol pipe union nut	19 Jet needle
9 Float chamber cover	20 Jet plug washer
10 Cable adjuster	21 Main jet
11 Mixing chamber top	22 Jet plug
12 Throttle spring	23 Float chamber

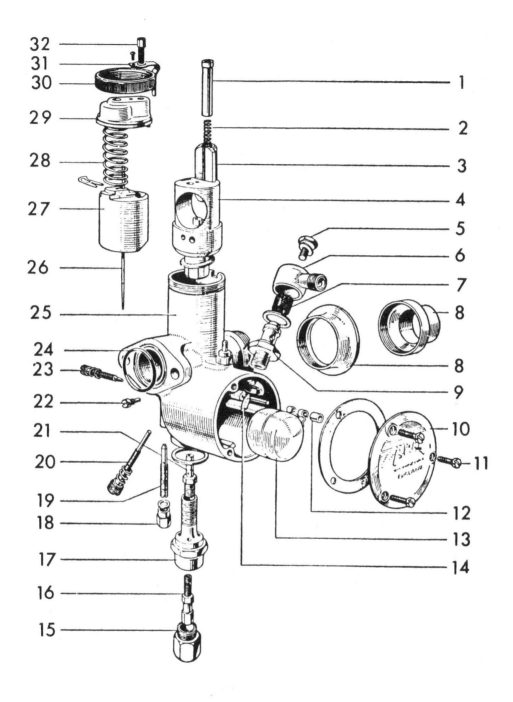

Fig.2.3. Component parts of the 'Monobloc' carburettor

1 Air valve guide	9 Needle setting	18 Pilot jet cover nut	27 Throttle slide
2 Air valve spring	10 Float chamber cover	19 Pilot jet	28 Throttle spring
3 Air valve	11 Cover screw	20 Throttle stop screw	29 Top
4 Jet block	12 Float spindle bush	21 Needle jet	30 Cap
5 Banjo bolt	13 Float	22 Locating peg	31 Click spring
6 Banjo	14 Float needle	231 Air screw	32 Adjuster
7 Filter gauze	15 Main jet cover	24 'O' ring seal	
8 Air filter connection (top)	16 Main jet	25 Mixing chamber	
or air intake tube	17 Main jet holder	26 Jet needle	

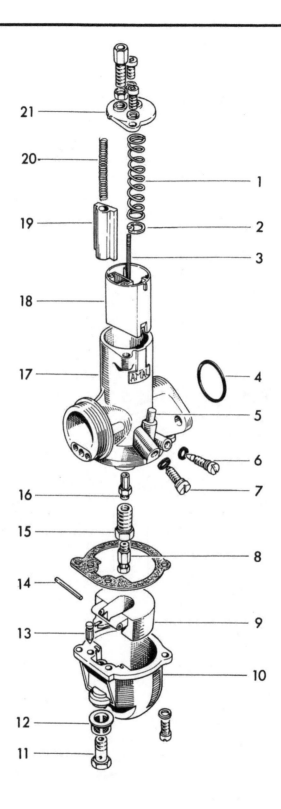

Fig.2.4. Component parts of the concentric carburettor

1 Throttle spring	7 Throttle stop	13 Float needle	19 Air slide
2 Needle clip	8 Main jet	14 Float spindle	20 Air slide spring
3 Throttle needle	9 Float	15 Jet holder	21 Mixing chamber cap
4 'O' ring	10 Float chamber body	16 Needle jet	
5 Tickler	11 Banjo bolt	17 Carburettor body	
6 Pilot air screw	12 Filter	18 Throttle valve	

intake by a rubber hose. A rubber retainer band seals and retains the air cleaner element to the back plate. When the element has been detached, it should be washed in petrol and allowed to dry, before replacement. No attention is required other than to repeat this procedure every 5,000 miles.

The "Bushman" model has a different type of air cleaner that is mounted on the sub-frame, below the seat. It is protected by a cover on the right-hand side of the machine, retained by a nut and bolt at the foremost end, a nut and bolt at the uppermost rear end and at its base by a mudguard bolt that needs slackening only. If the bolts securing the air cleaner bracket to the sub-frame clips are removed and the hose disconnected from the base plate, the complete unit can be withdrawn. The air cleaner contains a paper element, which is released when the centre bolt is removed from the top plate. If the element is dirty, it should be discarded and a new replacement fitted.

3 Do not run the machine with the air cleaner disconnected, particularly if the remote air cleaner unit is fitted. A smaller than standard main jet is fitted to the carburettor to compensate for the attachment of the air cleaner and if this jet is not changed, a very weak mixture will result, with the risk of consequent engine damage.

4 Check also that the connecting rubber hose is not split or damaged. An air leak in the hose will have a similar affect on carburation.

10 Crankcase Drain Plug - Location

1 A ¼ inch hexagon-headed drain plug is fitted in the base of the left-hand crankcase so that any excess oil can be drained off. It is also useful if the engine has been flooded with petrol, making starting difficult. The accumulation of petrol can be drained off when the plug is removed.

2 The plug must be tightened fully and have a fibre sealing washer. If it is loose or if there is an air leak, crankcase compression will be lost and engine performance will suffer accordingly.

3 Do not confuse the crankcase drain plug with the larger gearbox drain plug, located further back on the right-hand side.

11 Exhaust System- Cleaning

1 The exhaust system of any two-stroke engine requires quite frequent attention because the oily nature of the exhaust gases causes a build-up of sludge that will eventually partially block

the system and cause serious back-pressures. This will occur even more rapidly if the engine is in need of a rebore and is using oil.

2 The exhaust system is removed easily by following the procedure detailed in Chapter 1.5, Sections 3 - 4. The silencer separates from the exhaust pipe by unscrewing the pinch bolt of the clamp around the silencer/exhaust pipe joint.

3 The silencer fitted to the early D1 models does not have detachable baffles and it is necessary to fill the silencer with a solution of caustic soda after blocking up one end. If possible, leave the caustic soda solution within the silencer overnight before draining off and washing out thoroughly with water.

4 Caustic soda is highly corrosive and every care should be taken when mixing and handling the solution. Keep the solution away from the skin and more particularly the eyes. The wearing of rubber gloves is advised whilst the solution is being mixed and used.

5 The solution is prepared by adding 3 lbs of caustic soda to 1 gallon of COLD water, whilst stirring. Add the caustic soda a little at a time and NEVER add the water to the chemical. The solution will become hot during the mixing process, which is why cold water must be used.

6 Make sure the used caustic soda solution is disposed of safely, preferably by diluting with a large amount of water. Do not allow the solution to come into contact with aluminium castings, because it will react violently with this metal.

7 The later D1, the D and all subsequent models are fitted with a different type of silencer that has detachable baffles to aid cleaning. In this case it is not necessary to remove the exhaust system from the machine; the baffles can be removed from the silencer by unscrewing the nut at the end of the "fishtail" and removing the rod complete with baffles after the "fishtail" has been taken off. Note the "fishtail" is an aluminium alloy casting which must not be cleaned with caustic soda if the silencer complete is removed and this method of cleansing applied.

8 If the baffle assembly is heavily coated with a sludge of carbon and oil, it is permissible to burn this out with a blow lamp. Make sure the alloy "fishtail" is well out of range, or it may melt!

9 Never tamper with the exhaust system and remove the baffles from the silencer or fit a quite different system. Although a louder exhaust note may give the illusion of greater speed, in nearly every case performance will be reduced and the rider may risk prosecution. It is difficult to improve on the manufacturer's original specification, which has been designed to match in with the characteristics of the engine. Speed and noise do not necessarily go hand in hand.

11.7A. Fishtail is free when end nut is removed

11.7B. Rod withdraws complete with baffles

12 Fault Diagnosis - Fuel System and Carburation

Symptom	Cause	Remedy
Excessive fuel consumption	Air cleaner choked or restricted	Clean or if paper element fitted, replace.
	Fuel leaking from carburettor. Float sticking	Check all unions and gaskets. Float needle seat needs cleaning.
	Badly worn or distorted carburettor	Replace.
	Carburettor incorrectly adjusted	Tune and adjust as necessary.
	Incorrect silencer fitted to exhaust system	Do not deviate from manufacturer's original silencer design.
Idling speed too high	Throttle stop screw in too far. Carburettor top loose	Adjust screw. Tighten top.
Engine does not respond to throttle	Back pressure in silencer. Float displaced or punctured	Check baffles in silencer. Check whether float is correctly located or has petrol inside.
	Use of incorrect silencer or baffles missing	See above. Do not run without baffles.
Engine dies after running for a short while	Blocked air hole in filler cap	Clean.
	Dirt or water in carburettoror	Remove and clean out.
General lack of performance	Weak mixture; float needle stuck in seat	Remove float chamber or float and clean.
	Air leak at carburettor joint or in crankcase	Check joints to eliminate leakage.
Excessive white smoke from exhaust	Too much oil in petrol, or oil has separated out	Mix in recommended ratio only. Mix thoroughly if mixing pump not available.

Chapter 3 Ignition System

Contents

Specifications

D1 model - all engines up to engine number YD1—40660 (1948 — August 1950)
Flywheel Magneto Generator (a.c.)

Make and Type	Wico-Pacy 'Geni—Mag'
Output	Ignition and a.c. lighting (direct)
Contact breaker gap	0.015 inch

D1 model — all engines after engine number YD1—40661 and D3, D5 and D7 models
Flywheel Magneto Generator (a.c./d.c.)

Make and Type	Wico—Pacy Series 55/Mark 8
Output..	Ignition and rectified d.c. for lighting and battery charging
Contact breaker gap	0.015 inch

D1 model — Coil Ignition Version (1950—53)
Flywheel Alternator (a.c.)

Make and Type	Lucas 1A 45
Output..	Rectified d.c. for lighting and battery charging. Provision for Emergency start (flat battery)
Contact breaker gap	0.012 inch

D10 and D14 models
Flywheel Alternator (a.c.)

Make and Type	Wico—Pacy IG.1768
Output..	Rectified d.c. for lighting and battery charging. Provision for Emergency start (flat battery)
Contact breaker gap	0.012 inch

'Bushman' model
As above, but no rectifier or battery. Direct lighting

Sparking Plug

Make and Type –	Cast iron cylinder heads	Champion L—10	Lodge CC14 or CN
	Alloy cylinder heads...	Champion L—7	Lodge CC14 or HN
	D10 and D14 models, also 'Bushman' ...	Champion N—4	Lodge HLN

1 General Description

Several different types of ignition system are used on the BSA "Bantam" models and it may prove helpful if the sequence of changes is explained. The original D1 model employs a combined ignition and lighting system based on a flywheel magneto generator driven from the crankshaft. The generator contains both ignition and lighting coils and the ignition system is not dependent on a battery. It soon became apparent that some riders require a more powerful lighting system with a separate ignition coil to provide the spark. The circuit now

required a battery and also means of converting the a.c. output from the generator to d.c. so that the battery could be charged. These requirements were met initially by the Lucas 1A 45 alternation, which is fitted to all the early D1 battery lighting models. The Lucas ignition system is not wholly dependent on the battery; an Emergency Start procedure is provided so that the engine can be started when the battery is fully discharged, provided there is no additional electrical load. In due course the Lucas alternator was superceded by an improved Wico-Pacy flywheel magneto generator that has a much better spark output than the original, over a much wider range of engine speeds. There is also provision for including a battery in the lighting circuit, using the rectified output from the lighting coils to keep the battery charged. The Emergency Start procedure was no longer necessary; the machine can be run without the battery if the lights are not required. Eventually, a Wico-Pacy six-pole alternator was substituted for the Series 55 generator, as the result of further advances in the electrical industry. This latest system has reverted to the use of a separate ignition coil and battery, with provision for Emergency Start. The main advantages are higher output, better voltage regulation and a more reliable Emergency Start facility.

It is not possible to interchange parts from the different systems, even those originating from the same manufacture, in order to update the ignition and lighting equipment.

2 Flywheel Generators - Checking Output

1 The output and performance of the various types of generator fitted to the "Bantam" models can be checked only with specialised test equipment of the multi-meter type. It is unlikely that the average owner/rider will have access to this equipment or instruction in its use. In consequence, if the performance is suspect, the generator should be checked by a qualified auto-electrical expert.

2 Failure of the generator does not necessarily mean that a replacement is needed, unless this can be achieved economically through a service exchange scheme. It is possible to replace the individual ignition and/or lighting coils in the event of their failure, particularly in the earlier types of generator.

3 Ignition Coil - Checking

The ignition coil is a sealed unit, designed to give long service. Where a separate ignition coil is fitted, it is usually mounted in a position remote from the generator (behind the backplate of the left-hand sidecover, D10 and D14 models). If a weak spark and difficult starting cause its performance to be suspect, it should be tested by an auto-electrical expert. A faulty coil must be replaced; it is not practicable to effect a repair.

4 Contact Breaker - Adjustment

1 On the D1 and D3 models, access to the contact breaker assembly is gained by removing the circular cover plate on the left-hand generator cover. This is retained in position by either a spring clip or two cheese-headed screws, depending on the model. There is no cover plate on the D5 and D7 models; it is necessary to remove the outer generator cover complete, which is retained by three cheese headed screws, then the inner plate.

2 The D10 and D14 models have the contact breaker assembly mounted on the opposite (right-hand) side of the engine. Access is gained through a circular cover plate retained by two cross-head screws.

3 Rotate the engine until the contact breaker points are in the fully-open position. Examine the faces of the contacts. If they are pitted or burnt it will be necessary to remove them for further attention: as described in Section 4 of this Chapter.

4 Adjustment is carried out by slackening the screw that retains the fixed contact point and moving the plate upwards or downwards (adjust eccentric pin, D10 and D14 models), until the correct gap has been restored. Retighten the screw and check again that the gap is correct. It is important that the contacts are in the fully open position during this check, otherwise a completely false setting will be obtained. If the gap is correct, the feeler gauge should be a good sliding fit.

5 Before replacing the cover, place a very slight smear of grease on the contact breaker cam or one or two drips of light oil on the felt pad that presses on the cam (Wico-Pacy contacts).

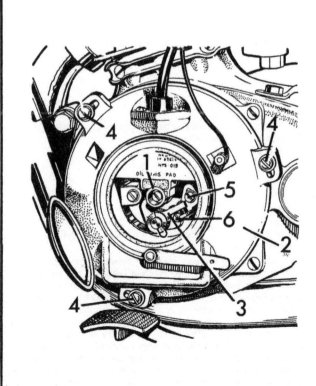

FIG.3.1. WICO-PACY 'GENI MAG' CONTACT BREAKER MECHANISM

1 Contact breaker cam
2 Housing
3 Rocker arm

4 Adjusting screws
5 Fixed contact screw
6 Terminal post

5 Contact Breaker Points - Removal, Renovation and Replacement

1 If the contact breaker points are burned, pitted or badly worn, they should be removed for dressing. If it is necessary to remove a substantial amount of material before the faces can be restored, the points should be replaced.

2 To remove the contact breaker points, slacken and remove the screw at the end of the moving contact return spring. Remove the spring clip and plain washer from the contact breaker arm pivot and withdraw the moving contact which is integral with the fibre rocker arm. Remove the screw that holds the fixed contact breaker point in position (also the eccentric pin, D10 and D14 models) and withdraw the plate complete with contact.

3 The points should be dressed with an oilstone or fine emery cloth. Keep them absolutely square during the dressing operation, otherwise they will make angular contact when they are replaced and will quickly burn away.

4 Replace the contacts by reversing the dismantling procedure. Take particular care when replacing the insulating washers, to make sure they are fitted in the correct order. If this precaution is not observed the points will be isolated electrically and the ignition system will not function.

5 The contact breaker assembly fitted to the D10 and D14 models differs in some respects from that of the earlier models, although the operating principle is the same. The same broad dismantling and reassembly procedure will apply. Do not omit the star-shaped washer on the moving contact breaker arm. It controls the end float and is essential to the correct functioning.

6 Condenser - Removal and Replacement

1 A condenser is included in the contact breaker circuitry to prevent arcing across the contact breaker points as they separate. It is connected in parallel with the points and if a fault develops the ignition system will not function correctly.

2 If the engine is difficult to start or if misfiring occurs, it is possible that the condenser has failed. To check, separate the contact breaker points by hand whilst the ignition is switched on. If a spark occurs across the points and they have a blackened or burnt appearance, the condenser can be regarded as unserviceable.

3 It is not possible to check the condenser without the necessary test equipment. In view of the low cost of a replacement it is preferable to change the condenser and observe the effect on engine performance.

4 To remove the condenser from the Wico-Pacy contact breaker withdraw the screw that clamps the condenser to the base plate of the stator assembly and detach the lead from the condenser where it joins the terminal post. The Lucas contact breaker has a similar arrangement. It is necessary to remove the stator plate assembly before access to the condenser can be gained in early models fitted with the Wico-Pacy "Geni-Mag".

FIG.3.2. WICO-PACY TYPE 55/MARK 8 CONTACT BREAKER MECHANISM

1 Contact breaker cam   4 Adjusting screws
2 Housing   5 Fixed contact screw
3 Rocker arm

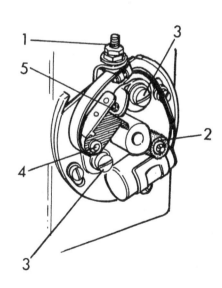

FIG.3.3. LUCAS 1A45 CONTACT BREAKER MECHANISM

1 Terminal post   plate contact
2 Felt lubricator   4 Contact breaker pivot
3 Screws securing fixed   5 Contacts

7 Ignition Timing - Checking and Resetting

1 If the ignition timing is correct, the contact breaker points will be about to separate when the piston is 5/32 in before top dead centre (TDC). This is the recommended setting for the D1 and D3 models. In the case of the D5 and D7 models, the piston should be only 1/16 in before top dead centre (TDC) when the contact breaker points are in the same position. A limited range of adjustment is provided by rotating the stator plate after the retaining screws have been slackened. (D1, D3, D5 and D7 models) or by rotating the contact breaker plate retained by four screws (early Lucas ignition models).

2 Optimum performance from the D10 and D14 models depends on very accurate setting of the ignition timing. A degree disc is essential, which should be attached to the generator end of the crankshaft after removing the generator outer cover (left-hand side) and the nut that retains the generator rotor.

3 A pointer should then be attached to one of the tapped holes used for the outer cover retaining screws and held firmly in position, so that it points to zero when the piston is exactly at top dead centre. Adjust by rotating the timing disc on the crankshaft and then lock the disc in position with the crankshaft rotor nut.

4 Disconnect and remove the battery from the machine. (A separate battery will be needed in the case of the "Bushman" model, to which a battery is not normally fitted). Attach one battery lead to the moving contact return spring and the other, via a 6 volt bulb, to any convenient earthing point on the machine. As soon as the contacts separate, the bulb will be extinguished.

5 Set the points so that the points are just separating and the bulb goes out. If the ignition timing is correct, the pointer should show a reading of $16\frac{1}{2}^{o}$ before top dead centre (TDC) on the degree disc.

6 A limited range of adjustment is available by turning the contact breaker plate, after loosening the cross-head retaining screws. The screw holes in the plate are slotted, to permit this movement. If it is still not possible to obtain an accurate setting, the contact breaker cam must be withdrawn and replaced in the correct position, by following the procedure given in Chapter 1.10 Sections 1 - 2.

7 Whatever method of timing is used, always recheck, especially if the timing has been altered. Even a comparatively small error will have a surprising effect on engine performance.

8 Note that the ignition timing is fixed. There is no provision for advancing the timing automatically as the engine speed increases.

8 Sparking Plug - Checking and Resetting Gap

1 A 14 mm sparking plug is fitted to all "Bantam" models, irrespective of whether the cylinder head is cast iron or alloy. Refer to the Specifications Section heading this Chapter for the recommended grades.

2 The D1, D3, D5 and D7 models use a sparking plug with ½ in reach, which should be gapped within the range 0.018 in - 0.020 in. The D10 and D14 models require a ¾ in reach plug. Always use the grade of plug recommended or the exact equivalent in another manufacturer's range.

3 Check the gap at the plug points every 1,000 miles. To reset the gap, bend the outer electrode closer to the central electrode and check that a 0.018 in feeler gauge can be inserted. Never bend the centre electrode, otherwise the insulator will crack, causing engine damage if particles fall in whilst the engine is running.

4 The condition of the sparking plug electrodes and insulator can be used as a reliable guide to engine operating conditions. See accompanying diagrams.

5 Always carry a spare sparking plug of the correct grade. The plug in a two-stroke engine leads a hard life and is liable to fail more readily than when fitted to its four-stroke counterpart.

6 Never over-tighten a sparking plug, otherwise there is risk of stripping the threads from the cylinder head, especially those cast in light alloy. A stripped thread can be repaired by using what is known as a 'Helicoil' insert, a low cost service that is operated by a number of dealers.

7 Use a spanner that is a good fit, otherwise the spanner may slip and break the plug insulator. The plug should be tightened sufficiently to seat firmly on its sealing washer.

8 Make sure the plug insulating cap is a good fit and free from cracks. This cap contains the suppressor that eliminates radio and tv interference.

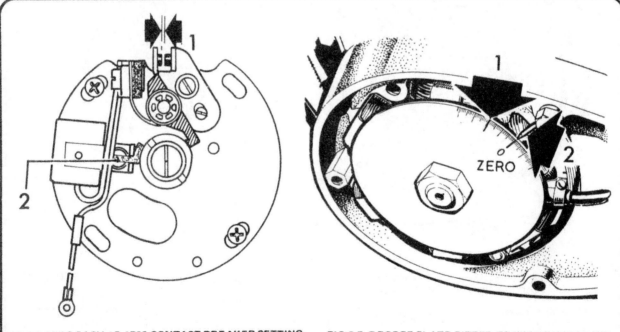

FIG.3.4. WICO-PACY 1G.1768 CONTACT BREAKER SETTING

    1 Contact points gap (0.012 in)   2 Felt pad

FIG.3.5. DEGREE PLATE FITTED TO WICO-PACY 1G.1768

    1 Timing angle on degree plate
    2 Pointer on crankcase (piston
    at top dead centre, TDC)

53

FIG.3.6. SAMPLE PLUG CONDITIONS

White deposits and damaged porcelain
insulation indicating overheating

Broken porcelain insulation
due to bent central electrode

Electrode burnt away due to wrong
heat value or chronic pre-ignition
(pinking)

Excessive black deposits caused
by over-rich mixture or wrong
heat value

Mild white deposits and electrode
burnt indicating too weak a fuel
mixture

Plug in sound condition with
light greyish-brown deposits

9 Fault Diagnosis - Ignition System

Symptom	Cause	Remedy
Engine will not start	No spark at plug	Try replacement plug if gap correct.
		Check whether contact breaker points are opening and closing, also whether they are clean.
		Check whether points arc when separated. If so, replace condenser.
		Check ignition switch (if fitted) and coil.
		Battery fully discharged. Switch off all lights and use emergency start (Lucas ignition models only).
Engine starts but runs erratically	Intermittent or weak spark	Try replacement plug. Check whether points are arcing. If so, replace condenser.
		Check accuracy of ignition timing.
		Low output from flywheel magneto generator or imminent breakdown of ignition coil.
		Plug has whiskered. Fit replacement.
		Plug lead insulation breaking down. Check for breaks in outer covering, particularly near frame.

Chapter 4 Frame and fork assembly

Contents

Specifications

D1 model (standard and competition)

Frame	Rigid or *plunger suspension (De Luxe model)
	Rigid only (competition)
Forks	Telescopic, undamped compression springs

D3 model (standard and competition)

Frame	Plunger suspension (1954—55)
	Swinging arm rear suspension (1956 onwards)
Forks	Telescopic, undamped compression springs

D5 and D7 models

Frame	Swinging arm rear suspension
Forks	Telescopic. D7 only has hydraulically—damped compression springs

D10 and D14 models

Frame	Swinging arm rear suspension
Forks	Telescopic, with hydraulically-damped compression springs

'Bushman'

Frame	As D14 models
Forks	As D14 models

Front forks (per leg):

Undamped type (D1, D3, D5)	Grease
Lightweight type (D7, D10, early D14/4)	70 — 75 cc ($\frac{1}{8}$Imp pt) of Castrolite
Heavyweight type (late D14/4, D175)	175 cc ($\frac{1}{3}$Imp pt) of Castrolite

*Production of the rigid frame D1 models ceased during 1955.

1 General Description

Three basic types of frame assembly are used for the "Bantam" models, a rigid frame, a frame with plunger-type rear suspension and a frame with swinging arm rear suspension. Initially, telescopic forks of the undamped compression spring type were fitted to all models.

The introduction of the 174 cc models necessitated some redesign, leading to an improved fork unit, hydraulic damping. It was not until the D10 models were announced that a completely new design of fork was fitted, which was continued through to the end of "Bantam" production. This latter type of fork is more sturdy and has a hydraulic damping system to give improved handling characteristics in keeping with the extended performance available from these newer models.

2 Front Forks - Removal from Frame

1 It is unlikely that the front forks will need to be removed from the frame as a complete unit, unless the steering head bearings require attention or the forks are damaged in an accident.

2 Commence operations by placing the machine on the centre stand and disconnecting the front brake. If the screw and nut through the "U" shaped connection to the brake operating arm are removed and the cable adjuster unscrewed from the brake plate, the operating cable can be pulled clear.

3 Remove the two front wheel nuts and washers, using the sparking plug spanner. If necessary, unscrew the three left-hand mudguard stay bolts, so that the left-hand fork leg can be raised and the front wheel dropped out at an angle. On the 174 cc

models it is necessary to remove the two bolts at the bottom of each fork leg, so that the split clamps can be separated and the wheel allowed to fall clear. It may be necessary to place a block under the centre stand so that the wheel is raised high enough for it to clear the front mudguard. The wheel will pull away complete with brake plate and spindle.

4 Disconnect the control cables from the handlebars by removing the controls complete or by disconnecting the cable ends. On the early models fitted with direct lighting, remove also the bulb horn that passes through the steering head and is held in position by a locknut. The bulb will unscrew from the main body of the horn, to permit access to this nut.

5 Remove the handlebars by separating the split clamps, or in the case of the D3, D5 and D7 models, the aluminium cover that acts as a clamp. Disconnect the speedometer cable by unscrewing the connection at the bottom of the speedometer head and pulling the cable free. Unscrew also the bulb holder of the speedometer illuminating lamp.

6 Remove the headlamp with the remote control switch still attached (early models) after making note of any of the internal connections if it is necessary to detach additional wires. On the D5 and later models this task is much easier because the ignition switch and lighting switch cable sockets can be detached from the top of the headlamp nacelle. They are a push fit and are retained in position by a spring clip. If the large rubber grommet

at the base of the nacelle is then displaced, the cable harness complete with sockets can be withdrawn.

7 Slacken the pinch bolt of the fork top yoke, located to the rear of the steering head stem. If necessary, remove the bolt and spring open the gap.

8 Unscrew the fork leg nuts, at the top of each fork leg, and remove them complete with washers. The D7 models have an additional spring cap within the fork leg nut that must be removed before the inner fork spring nuts can be slackened and withdrawn and the fork leg nuts removed. A very slim socket or box spanner is needed for these nuts. Slacken and remove the large nut at the top of the steering head. The D3, D5 and D7 models have a variation of this arrangement in which a dust cover fulfills the function of the steering head nut. When this cover is removed, a slotted adjuster will be found, which is a push fit into the top yoke.

9 A few light taps with a rawhide mallet in an upwards direction will displace the top yoke and allow the complete fork unit to be withdrawn from the bottom of the steering head with the bottom yoke and head stem. Make provision for catching the uncaged ball bearings that will be released as the cup and cone steering head bearings are separated.

10 If further dismantling is necessary, the front mudguard can be removed when the forks have been withdrawn, by unscrewing the lower connections to the mudguard stays.

2.8A. Spring caps lift out on D7 models

2.8B. Slim socket spanner is needed for fork spring nuts due to minimum clearance

2.10A. Lower mudguard stay fixings on D7 models act also as drain plugs

2.10B. Remove mudguard before dismantling fork legs

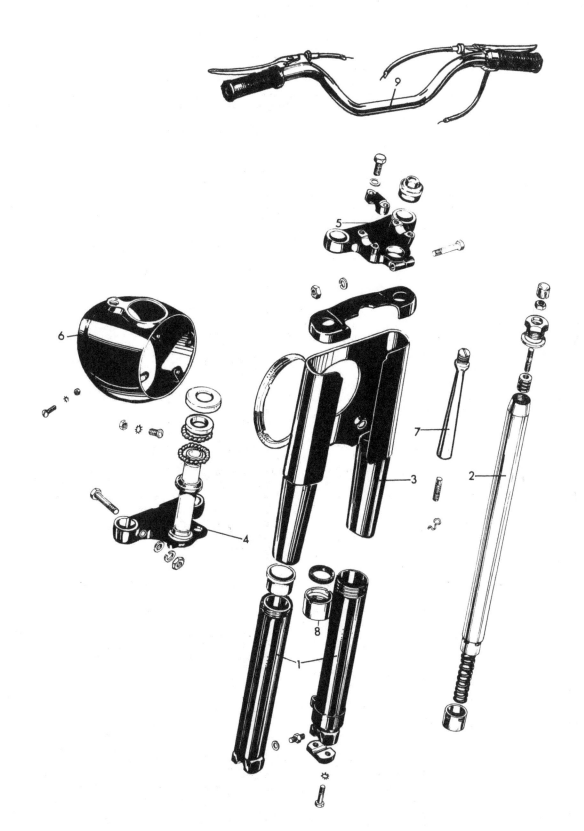

FIG.4.1. FRONT FORK COMPONENTS

1 Lower fork legs	3 Fork shrouds	6 Headlamp nacelle	8 Oil seal cup
2 Upper fork legs	4 Bottom yoke assembly	7 Restrictor (fork	9 Handlebars and controls
	5 Top yoke	damper)	

3   Front Forks - Dismantling

1   To remove the fork legs, unscrew the pinch bolts in the fork bottom yoke. The fork legs should now pull clear. If they are still a tight fit, spring open the pinch bolt joint a little. In the case of hydraulically—damped forks fitted to the D7, D10 D14 models, it is preferable to drain off the oil content of each leg before they are withdrawn from the bottom yoke. A drain plug will be found at the base of each leg, close to the front wheel spindle clamps. The forks fitted to the D7 models have extended drain plugs that provide also the fixings for the lower mudguard stays. The stays must be detached before the drain plugs can be withdrawn.

2   Fork legs of the undamped type (D1, D3, D5 and D7 models) are dismantled by unscrewing the plated cup at the lower end of each of the fork tubes. The cup has a normal right-hand thread. When the cup has been unscrewed the lower sliding member can be drawn down and out of the fork tube. complete with spring. The D7 forks, although hydraulically-damped, dismantle in the same way.

3   To unscrew the spring, hold the lower fork leg in a vice, taking care not to damage the screw threads, and remove the spring by tapping the end with a hammer and small punch.

4   The hydraulically-damped fork legs of the D10 and D14 models are separated from the fork bottom yoke in an identical manner, but a revised dismantling procedure is then necessary, due to their different construction.

5   Pull the fork spring out of the oil seal holder, then hold the lower fork leg in a vice so that the oil seal holder can be removed by unscrewing it in an anti-clockwise direction. The holder has two peg holes in its outer surface, which locate with the pegs of BSA Service Tool 61-6017 that is specified for this task. If the Service Tool is not available, skillful use of a centre punch should provide a satisfactory alternative. Do not use undue force or the body of the holder will be distorted and no longer provide an effective location for the bottom oil seal.

6   The main fork tube can now be detached from the lower fork leg by jerking it upwards, to free the spacer within the bottom fork leg.

4   Front Forks - General Examination

1   Apart from the oil seals and bushes, it is unlikely that the forks will require any additional attention unless the fork springs are weak or if the fork legs or yokes have been damaged in an accident. The undamped forks fitted to the D1, D3 and D5 models should have a maximum travel of 3¾ in; if this is exceeded, it is probable that the fork springs have weakened and require replacing. The free length of the springs fitted to the hydraulically-damped forks should be 10.5/32 inch.

2   Visual examination will show whether the fork yokes are distorted or if the inner fork tubes are bent (D7, D10 and D14 models). It is rarely possible to effect a satisfactory repair and replacement is strongly recommended.

5   Front forks – Examination and Replacement of Oil Seals

1   It is good practice to renew the oil seals during front fork overhaul, irrespective of whether they have shown signs of leakage. The seals are housed in the oil seal holders, which must first be removed as described in Section 3.

2   On the D1, D3 and D5 fork, the double-lipped grease seal is simply a press fit in its holder. The D7 and D10 models are fitted with a short seal holder, in which the unsprung oil seal is a press fit. On D14/4 models the oil seal is retained by a circlip in its seal holder. A tall seal holder with chrome ring on the outside is fitted to D14/4S models; the seal is a drive fit in the holder. On the Heavyweight type forks fitted to the D14/4B, D175 and B175 models, the tall seal holder contains a wiper

3.1. Pinch bolts retain fork legs (applies to D7 models only)

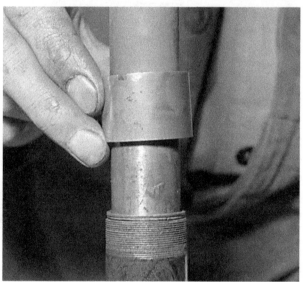

3.2. Plated cups allow lower fork leg to be withdrawn when they are unscrewed

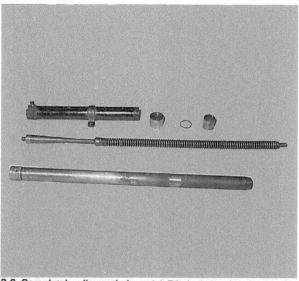

3.6. Completely dismantled model D7 fork leg (applies to D7 models only)

seal located above the oil seal and a washer and O-ring situated immediately beneath it. It is recommended that both the wiper seal and O-ring also be renewed during overhaul.

3 On the early models it should be possible to simply press the seal from position, but on later models where the seal is a drive-fit (D14/4S, D14/4B, D175 and B175 models), it must be driven from position. Place the seal holder upside down on a block of wood. Using a suitably sized drift work around the seal until it can be withdrawn from the top of its holder. When fitting the new seal, smear jointing compound over its outside edge and drive into position. Move the drift around the face of the seal during fitting to ensure that it enters its holder squarely. On all models smear a trace of grease over the seal lips prior to fork reassembly.

6 Front Forks - Examination and Replacement of Bushes

1 Some indication of the extent of wear of the fork bushes can be gained before the machine is dismantled. If the front wheel is gripped between the knees and the handlebars rocked to and fro, the amount of wear will be magnified by the leverage at the handlebar ends. Cross-check by applying the front brake and pulling and pushing the machine backwards and forwards.

2 The fork bushes are located within outer fork tubes and in the case of machines manufactured prior to May 1951, they cannot be replaced. If the bushes show signs of wear, the outer fork tubes complete must be replaced. Although it is possible to remove and replace the bushes in the later D1, D3 and D5 models, this is not an easy task and BSA Motor Cycles have always recommended fitting replacement outer tubes.

3 The fork bushes of the hydraulically-damped forks (D10 and D14 models) are located within the lower fork legs and also at the end of each main fork tube. The former are a light press fit and can be removed and replaced without difficulty. The latter are held captive by the restrictor nuts, which must be unscrewed (15/16 in spanner across flats) before they can be withdrawn. Take note of any washers and shims in the assembly, which must be replaced. The D7 forks have a somewhat similar arrangement although a simplified damping system is used, with fewer parts.

7 Steering Head Bearings - Examination and Replacement

1 Before commencing to reassemble the forks, inspect the steering head races. The ball bearing tracks should be polished and free from indentations and cracks. If signs of wear or damage are evident, the cups and cones must be drifted out of position. They are a tight press fit and need to be drifted out of position.

2 Ball bearings are cheap. Each race contains twenty four 3/16 in. balls, which should be replaced without question if the originals are marked or discoloured. To hold the ball bearings in place whilst the forks are reattached, pack the bearings with grease.

8 Front Forks - Reassembly

1 To reassemble the front forks, follow the dismantling procedure in reverse. Take particular care when passing the sliding fork members through the oil seals because the seals are very easily damaged. It is advisable to smear the sliding members with grease as well as the inside lips of each seal.

2 Tighten the steering head carefully, so that all play is eliminated without placing undue stress on the bearings. The adjustment is correct if all play is eliminated and the handlebars will swing to full lock of their own accord when given a push on one end.

3 It is possible to place several tons pressure on the steering head bearings if they are over-tightened. The usual symptom of over-tight bearings is a tendency for the machine to roll at low speeds, even though the handlebars may appear to turn quite freely.

4 If after assembly it is found that the forks are incorrectly aligned or unduly stiff in action, loosen the front wheel spindle, the two top fork leg nuts, and the pinch bolts in both the top and bottom yokes. The forks should then be pumped up and down several times to realign them. Retighten all the nuts and bolts in the same order, finishing with the steering head pinch bolt.

5 This same procedure can be adopted if the forks are misaligned after an accident. Often the legs will twist within the fork yokes giving the impression of more serious damage, even though no structural damage has occurred.

6 Do not forget to add the recommended quantity of oil to each fork leg before replacing the top fork leg nuts of the D7, D10 and D14 models.

9 Frame Assembly - Examination and Renovation

1 A rigid frame is used for the early D1 models, which should not require attention unless it is damaged in an accident. Frame repairs are best entrusted to a specialist in this type of work, who will have all the necessary jigs and mandrels available. In many instances, a replacement frame from a breaker's yard is the cheaper and more satisfactory alternative.

2 If the machine is stripped for an overhaul, this affords an excellent opportunity to inspect the frame for any cracks or other damage that may have occurred in service. Check the front down tube at the point immediately below the steering head, which is where a break is most likely to occur. Check also the top tube of the frame, for straightness.

3 The early D1 spring frame models employ what is known as the plunger form of rear suspension in which the vertical movement of the rear wheel is controlled by undamped coil springs. Wear is likely to occur in the bushes of the sliding member and it is necessary to dismantle each suspension unit to gain access.

4 Remove the rear wheel, as detailed in Chapter 5, Section 6. Disconnect the mudguard stays and remove the pinch bolts at the top and bottom of each suspension unit. The centre column can now be drifted out, using a soft drift to avoid damage to the end of the column.

5 Grip the upper and lower shrouds with both hands to compress the internal springs. This will permit the complete suspension unit to be withdrawn from the frame mountings.

6 Remove the shrouds and springs from the sliding member, noting the position of the steel washers and rubber bushes for reassembly. The bushes in the sliding member can now be examined for wear; should they require renewing, the tube complete with bushes must be replaced. To withdraw the tube, unscrew the pinch bolt and open the slot with the blade of a screwdriver.

7 Replace the units by following the dismantling procedure in reverse. When replacing the sliding member, note that the hole in the side must line up with the grease nipple in the fork end. Place the lower shroud in the bottom frame lug and press down on the upper shroud to compress the springs. It is essential that the units are replaced in the correct order because the nearside fork end carries the anchor lug for the rear brake plate.

8 The D5, D7, D10 and D14 models employ swinging arm rear suspension in which the rear subframe pivots from a point to the rear of the gearbox. Movement is controlled by two hydraulically-damped rear suspension units. More detailed attention is required when this form of rear springing requires inspection and it is necessary to devote the following complete Section to this quite different mode of rear suspension.

10 Swinging Arm Rear Suspension - Examination and Renovation

1 After an extended period of service, the bush and pivot pin of the swinging arm fork will wear, giving rise to lateral play that will affect the handling characteristics of the machine. The bush is a tight press fit in the main frame and cannot be removed without risk of damage unless the correct equipment is available.

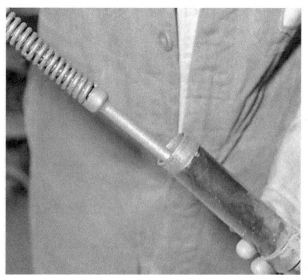

8.1A. Reconnect end of spring assembly with lower fork leg (applies to D7 models only)

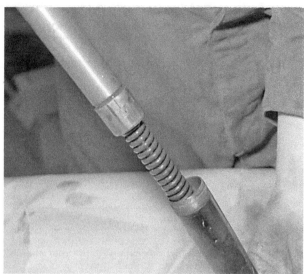

8.1B. Insert fork leg in lower tube, over spring assembly (applies to D7 models only)...

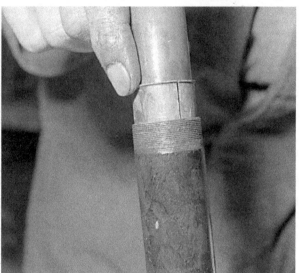

8.1C. ...then insert split bush, followed by... (applies to D7 models only)

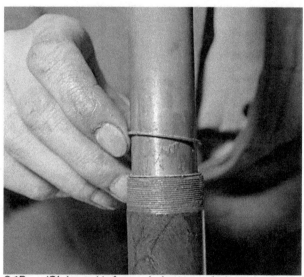

8.1D. ...'O' ring seal before replacing screwed cup (applies to D7 models only)

8.1E. Screwdriver in pinch bolt joint makes reassembly of fork tubes easier

8.6. Oil must be added to restore damping action

This form of repair is best entrusted to a BSA repair specialist, who will have the appropriate equipment. It is highly improbable that the average owner/rider will have access to this equipment or the skill with which to undertake the reconditioning work necessary.

2 The rear suspension units are removed by taking out the top fixing bolt and nut and washers and the lower fixing nut.

11 Rear Suspension Units - Examination

1 Only a limited amount of dismantling can be undertaken because the damper unit is sealed and cannot be replaced. If the unit leaks oil, or if the damping action is lost, the unit must be replaced as a whole or after removing the compression spring and shroud.

2 Before the shroud and spring can be removed, it is necessary to displace the split collets inside the base of the shroud. The task is made easier if BSA Service Tool 61-5064 is used; there is less of a problem with the "Bushman" models that have no shroud around the spring.

12 Centre Stand - Examination

1 The centre stand is attached to a lug on the bottom frame tube, to provide a convenient means of parking the machine on level ground. It pivots on a long bolt that passes through the lug, secured by a nut and washer. A return spring retracts the stand when the machine is pushed forward, so that it can be wheeled prior to riding.

2 The condition of the return spring and the return action should be checked regularly, also the security of the retaining nut and bolt. If the stand stops whilst the machine is in motion it could catch in some obstacle in the road and unseat the rider.

13 Footrests - Examination and Renovation

1 The footrests are attached to the same bottom frame tube lug as the centre stand, using a long bolt and nut. The footrests are malleable and will bend if the machine is dropped.

2 To straighten the footrests, first remove them from the machine and clamp them in a vice. They should be bent straight by applying leverage from a long tube that slips over the end, whilst heating the area in which the bend occurs to a cherry red with a blow lamp.

14 Speedometer - Removal and Replacement

1 The early D1 and D3 models are fitted with a 'D' shaped Smiths Chronometric speedometer, calibrated up to 65 mph. Some have an internal lamp for illuminating the dial. Later models are fitted with a Smiths magnetic speedometer, calibrated up to 85 mph. There is no internal bulb for illuminating the dial.

2 The 'D' shaped speedometer heads have two studs, which pass through the fork top yoke to act as a means of attachment. The magnetic speedometer heads are recessed into the headlamp nacelle and retained in position by a simple clamp arrangement. If the nuts are unscrewed and the clamp removed, the complete speedometer head can be lifted clear from the top, after the speedometer drive cable has been detached.

3 Apart from defects in either the speedometer drive or the drive cable itself, a speedometer that malfunctions is difficult to repair. Fit a replacement or alternatively entrust the repair to an instrument repair specialist, bearing in mind that a speedometer that functions in an efficient manner is a statutory requirement.

4 Each type of speedometer head fitted contains a continuously recording odometer, as a check on the mileage covered. There is no means of resetting the odometer reading.

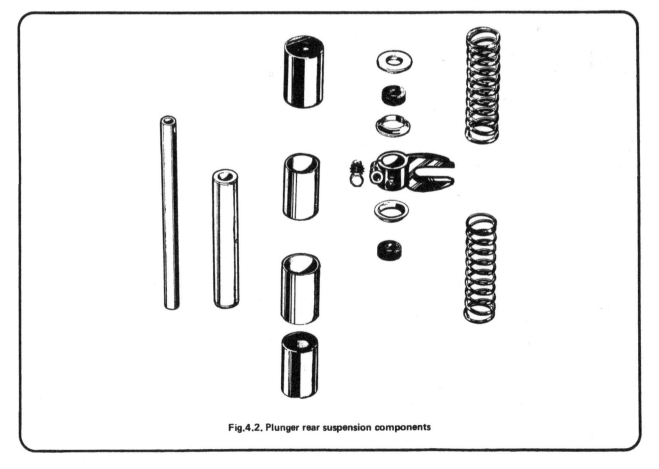

Fig.4.2. Plunger rear suspension components

15 Speedometer Cable - Examination and Renovation

1 It is advisable to detach the speedometer drive cable from time to time, in order to check whether it is adequately lubricated and whether the outer covering is compressed or damaged at any point along its run. A jerky or sluggish speedometer movement can often be attributed to a cable fault.

2 To grease the cable, withdraw the inner cable. After removing the old grease, clean with a petrol soaked rag and examine the cable for broken strands or other damage.

3 Regrease the cable with high melting point grease, taking care not to grease the last six inches at the point where the cable enters the speedometer head. If this precaution is not observed, grease will work into the speedometer head and immobilise the movement.

4 If both the speedometer head and the odometer stop working, it is probable that the speedometer cable has broken. Inspection will show whether the inner cable has broken, if so, the inner cable alone can be renewed and reinserted in the outer covering, after greasing. Never fit a new inner cable alone if the outer cover is damaged or compressed at any point along its run.

16 Dualseat - Removal

1 To remove the dualseat, first loosen the fixing bolts of the rear suspension units so that the ends of the rear dualseat bracket can be pulled free.

2 Raise the seat at the rear and pull backwards until the front clip disengages from the attachment on the top frame tube.

3 Early D1 models are fitted with a saddle. This can be removed by withdrawing the nut and bolt underneath the nose of the saddle and the nuts and bolts through the two rear-mounted springs.

17 Petrol Tank Badges - Replacement

1 The D10 and D14 models have circular plastics tank badges, mounted within a recess on each side of the petrol tank. These badges should not require replacement unless the machine is dropped and they are cracked as a result.

2 To remove a badge, withdraw the two slotted head screws that retain it in position. This will release the clip that holds the badge to the tank, through a hasp welded within the recess. The badge can be replaced only when the clip is positively located.

18 Steering Head Lock

1 Commencing in 1956 all models are fitted with a steering head lock. When the lock is actuated, a tongue protrudes through a hole in an extension of the steering head base, to secure the handlebars on full left lock. No attention is necessary other than the occasional application of light oil. If the lock malfunctions, it must be replaced.

2 Later models have a simplified arrangement, in which two holes coincide when the forks are on full left lock. The holes are large enough to accept the hasp of a padlock.

19 Prop Stand

1 The "Bushman" models have a prop stand in place of the centre stand fitted to the standard models. The same advice as given in Section 11 of this Chapter applies to the maintenance of this fitting.

FIG.4.3. FRAME AND SWING ARM COMPONENTS

1	Frame	7 Footrest assembly
2	Swinging arm	8 Pillion footrests
3	Rear suspension units	9 Rear brake rod
4	Sub-frame/rear mud-guard stay	10 Swinging arm pivot pin
5	Engine plates	11 Swinging arm bush
6	Rear brake pedal	12 Centre stand assembly
		13 Prop stand

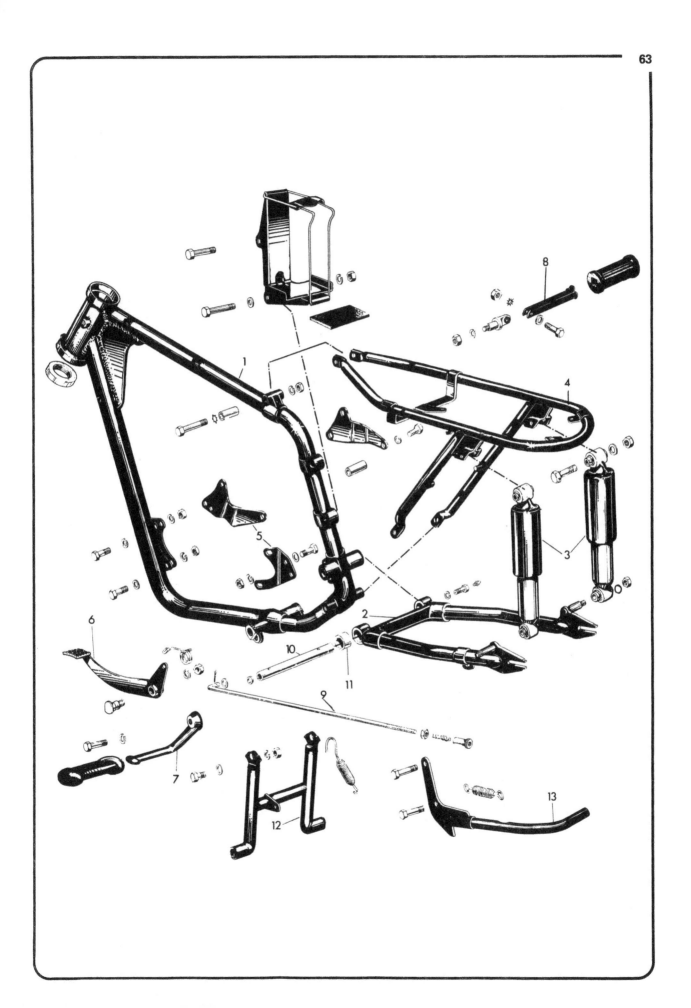

20 Fault Diagnosis - Frame and Fork Assembly

Symptom	Cause	Remedy
Machine is unduly sensitive to road surface irregularities	Fork and/or rear suspension units damping ineffective	Check oil level in forks. Replace rear suspension units.
Machine rolls at low speeds	Steering head bearings overtight or damaged	Slacken bearing adjustment. If no improvement, dismantle and inspect head bearings.
Machine tends to wander; steering is imprecise	Worn swinging arm suspension bearings	Check and if necessary renew pivot spindle and bush.
Fork action stiff	Fork legs have twisted in yokes or have been drawn together at lower ends	Slacken off spindle nut (clamps), pinch bolts in yokes and fork top nuts. Pump forks several times before re-tightening from bottom. Add distance pieces to fork spindle (early models).
Forks judder when front brake is applied	Worn fork bushes	Strip forks and replace worn bushes.
	Steering head bearings too slack	Re-adjust to take up play.
Wheels seem out of alignment	Frame distorted as result of accident damage	Check frame after stripping out. If bent, specialist repair is necessary.

Chapter 5 Wheels, brakes and transmission

Contents

Specifications

Wheels
(D1, D3 and D5 models) ... ... 19 inch diameter, front and rear
(D7, D10 and D14 models)... ... ... 18 inch diameter, front and rear
("Bushman" models).. ... ... 19 inch diameter, front and 18 inch diameter rear

Tyres
(D1, D3 and D5 models) ... ... 2.75 inch x 19 inch, front and rear
(D1 and D3 Competition models) ... ... 2.75 inch x 19 inch, front and 3.25 inch x 19 inch rear
(D7, D10 and D14 models)... ... ... 3.00 inch x 18 inch, front and rear
("Bushman" models).. ... ... 3.00 inch x 19 inch front and 3.25 inch x 18 inch rear

Brakes
(D1, D3 and D5 models) ... ... 5 inch diameter, front and rear
(D7, D10, D14 and "Bushman" models) ... ... 5½ inch diameter, front and rear

1 General Description

The D1, D3 and D5 models have 19 in diameter wheels, fitted with tyres of 2.75 in section. The Competition models retain the same section front tyre, but have a 3.25 in tyre fitted to the rear wheel. With the introduction of the D7 model, the wheel sizes were reduced to 18 in diameter and 3.00 in section tyres are fitted as standard. These sizes continued through to the end of production, the only exception being the "Bushman" model which uses a 19 in front wheel fitted with a 3.00 in section tyre and an 18 in rear wheel fitted with a 3.25 in section tyre. Initially, 5 in diameter brakes were fitted to the D1, D3 and D5 models. The D7 and later models employ 5½ inch diameter brakes, in order to keep the braking characteristics in line with the improved performance available. The rear wheel is not quickly detachable on any of the "Bantam" models; the chain must be detached before the wheel can be removed from the frame.

2 Front Wheel - Examination and Renovation

1 Place the machine on the centre stand so that the front wheel is raised clear of the ground. Spin the wheel and check for rim alignment. Small irregularities can be corrected by tightening the spokes in the affected area, although a certain amount of experience is necessary if over-correction is to be avoided. Any 'flats' in the wheel rim should be evident at the same time. These are more difficult to remove with any success and in most cases the wheel will need to be rebuilt on a new rim. Apart from the effect on stability, there is greater risk of damage to the tyre bead and walls if the machine is run with a deformed wheel.

2 Check for loose or broken spokes. Tapping the spokes is the best guide to tension. A loose spoke will produce a quite different note and should be tightened by turning the nipple in an anti-clockwise direction. Always check for run-out by spinning the wheel again.

3 Front Brake Assembly - Examination, Renovation and Reassembly

1 The front brake assembly complete with brake plate can be withdrawn from the front wheel by following the procedure in Chapter 4, Section 2. In the case of the D7, D10 and D14 models the centre nut that locks the brake plate to the spindle, must also be removed before the brake assembly can be withdrawn.

2 Examine the condition of the brake linings. If they are wearing thin or unevenly, the brake shoes should be replaced.

3 To remove the brake shoes from the brake plate, pull them apart whilst lifting them upwards in the form of a 'V'. When they are clear of the brake plate, the return springs can be removed and the shoes separated.

4 It is possible to replace the brake linings on the shoes fitted to the early D1 and D3 models, where the linings are rivetted to the brake shoe and not bonded on, as in current practice. Much will depend on the availability of the original type of linings; bonded on linings complete with replacement shoes may be the only practical form of replacement.

5 Before replacing the brake shoes, check that the brake operating cam is working smoothly and not binding in its pivot. The cam can be removed for greasing by unscrewing the nut on the brake operating arm and drawing the arm off, so that the spindle and cam can be pushed out of the housing in the back plate.

6 Check also the inner surface of the brake drum, on which the brake shoes bear. The surface should be smooth and free from score marks or indentations, otherwise reducing braking efficiency is inevitable. Remove all traces of brake lining dust and wipe with a rag soaked in petrol to remove any traces of grease or oil.

7 To reassemble the brake shoes on the brake plate, fit the return springs and force the shoes apart, holding them in a 'V' formation. If they are now located with the brake operating cam and pivot they can usually be snapped into position by pressing downwards. Do not use excessive force, or the shoes may be distorted permanently.

4 Wheel Bearings - Examination and Replacement

1 When the brake plate complete with brake assembly has been removed, the bearing retainer inside the brake drum will be exposed. This has a left-hand thread and is removed by using either a peg spanner (BSA Service Tool 61-3644) or a centre punch. The bearing is extracted after the retainer has been removed by striking the left-hand end of the front wheel spindle, with a rawhide mallet. A shim is inserted between the bearing and the shoulder of the wheel spindle, which must not be omitted during reassembly.

2 The left-hand bearing is removed by detaching the retaining circlip, then inserting the wheel spindle from the right-hand side of the hub, using it to drift the bearing outwards. The bearing will be preceded by a dust cover.

3 The D1 and D3 Competition models employ taper roller bearings in the front wheel hub, necessitating a revised dismantling procedure. If the adjusting locknuts are removed from the left-hand side of the front wheel spindle, the spindle can be withdrawn from the right-hand side by giving a light tap on the end. The inner and outer races can then be separated.

4 Remove all the old grease from the hub and bearings, giving the latter a final wash in petrol. Check the bearings for play or signs of roughness when they are turned. If there is any doubt about their condition, play safe and replace them.

5 Before replacing the bearings, first pack the hub with new grease. Then grease both bearings and drive them back into position, using either a double-diameter drift or the wheel spindle. Make sure the bearing retainer and the dust cover are located correctly, also the various distance pieces. The bearing retainer performs also the function of preventing grease from entering the brake drum and reducing the efficiency of the brake.

6 The taper roller bearings fitted to the early Competition models will require adjustment after reassembly. The amount of play is taken up by the two locknuts, which should be arranged so that the wheel will rotate freely with just perceptible play at the wheel rim. Make sure the locknuts are tightened fully after the adjustment has been completed, or they may work loose as the wheel rotates. Carry out a final check for play at the wheel rim.

7 After lengthy service and a number of adjustments to the taper roller bearings, the forks may show a tendency to bind because the fork ends are being pulled closer together. It is permissible to compensate for this misalignment by adding shims between the locknuts and the left-hand fork end to match the amount taken up by the adjustments.

5 Front Wheel - Reassembly and Replacement

1 Place the front brake assembly in the brake drum and align the front wheel so that the hole in the brake plate engages with the peg on the lower right-hand fork leg. This is important because the anchorage of the front brake plate is dependent on the correct location of these parts.

2 Reconnect the front brake and check that the brake functions correctly, especially if the adjustment has been altered or the brake operating arm has been removed and replaced during the dismantling operation.

6 Rear Wheel - Examination, Removal and Renovation

1 Before removing the rear wheel, check for rim alignment, damage to the rim and loose or broken spoked by following the procedure that applies to the front wheel, in the preceding Section.

2 To remove the rear wheel, place the machine on the centre stand so that the wheel is raised clear of the ground. Disconnect the rear chain at its spring link, a task that is made easier if the spring link is first positioned so that it is on the rear wheel sprocket. Unwind the chain off the sprocket and lay it on a clean surface.

3 Take off the rear brake rod adjuster and pull the brake rod and spring clear from the brake operating arm.

4 Remove the nut and bolt that attach the brake plate torque arm to the swinging arm fork (D7, D10 and D14 models only).

5 Disconnect the speedometer drive cable from the gearbox drive on the right-hand side of the wheel spindle.

6 Unscrew the two rear spindle nuts and pull the rear wheel from the fork ends. The wheel will pull out complete with the chain adjusters and the speedometer gearbox. It may be necessary to lean the machine slightly to the left to give sufficient clearance for the wheel to be removed from the right.

7 Avoid disturbing the chain adjusters since they determine the way in which the wheel is located with respect to the swinging arm fork.

8 The speedometer gearbox is held in position by a locknut. If this nut is removed, the gearbox can be withdrawn from the hub as a complete unit. Apply the brake to prevent the spindle from turning whilst the nut is slackened. The rear brake plate is retained by a detachable distance piece over the spindle (locknut on D7, D10 and D14 models).

9 The rear wheel bearings are of the non-adjustable journal ball type. They are a drive fit in the hub and can be removed by following the procedure recommended for the removal of the front wheel bearings, in Section 4. Note that in this instance, the bearing retainer within the brake drum has a normal right hand thread as the brake drum is on the opposite side.

10 Make sure the distance piece and the driving dogs are not misplaced when the speedometer drive gearbox is removed from the wheel spindle. They may fall free quite readily.

3.7. Snap raised brake shoe back into position to complete re-assembly

5.1. Some models have a torque anchorage formed by a strap around a fork leg

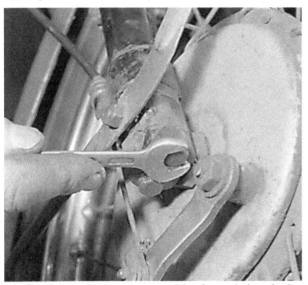

5.2. Check that split clamp bolts are tight after replacing wheel

6.4. Remove bolt from rear brake torque arm when detaching wheel

6.5. Disconnect speedometer drive cable from gearbox drive

6.6. Lean machine to right or left to gain clearance for wheel removal

6.7. Chain adjusters need not be disturbed

6.8A. Speedometer drive gearbox is retained by nut

6.8B. Brake plate assembly is also retained by nut

6.9A. Bearings drive out of hub

6.9B. Bearing retainer has right hand thread

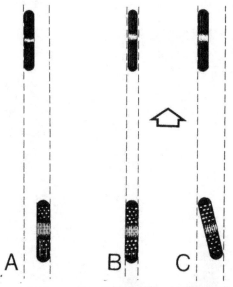

A   B   C

FIG.5.1. CHECKING WHEEL ALIGNMENT

A and C — Indicate necessity to re-align rear wheel
B —   Indicate correct alignment with smaller section front tyre

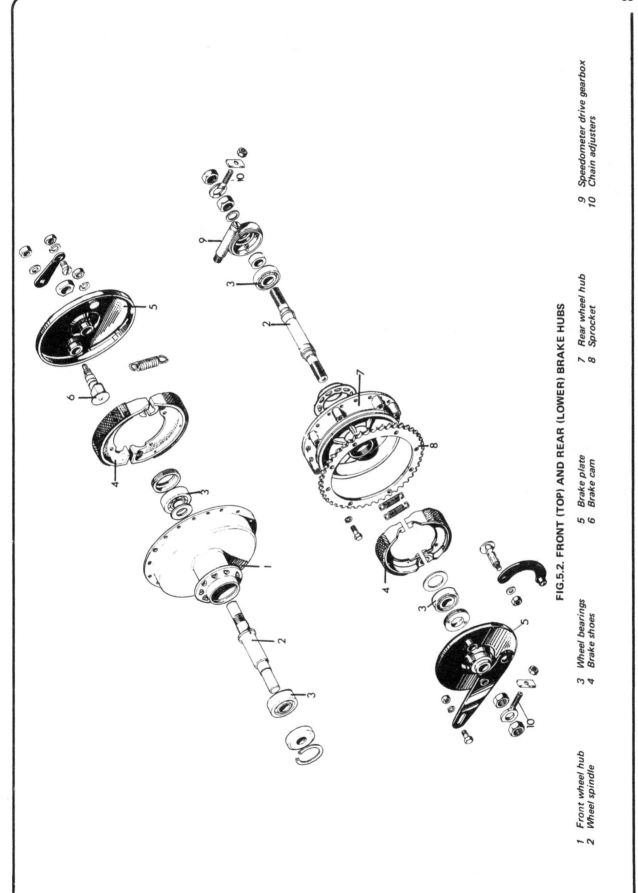

FIG.5.2. FRONT (TOP) AND REAR (LOWER) BRAKE HUBS

1	Front wheel hub	3	Wheel bearings	5	Brake plate	7	Rear wheel hub	9	Speedometer drive gearbox
2	Wheel spindle	4	Brake shoes	6	Brake cam	8	Sprocket	10	Chain adjusters

7 Rear Brake Assembly - Examination, Renovation and Reassembly

1 The rear brake assembly can be withdrawn from the rear wheel, complete with brake plate after the wheel has been removed from the frame and the distance piece or locknut removed. Refer to the preceding Section of this Chapter for details.

2 Follow an identical procedure for dismantling and reassembly to that relating to the front brake, as described in Section 3 of this Chapter. The brake assemblies are identical, even to the extent of interchangeable brake plates on the early D1 and D3 models.

8 Rear Wheel Bearings - Examination and Replacement

1 The rear wheel bearings are of the journal ball type, similar to those fitted to the front wheel. Replacement is necessary if play is evident or if there is any roughness when the bearings are turned, after they have been washed out with petrol.

2 The greasing and reassembly procedure is identical to that described in Section 4, paragraphs 4 and 5.

9 Front and Rear Brakes - Adjustment

1 The front brake adjuster is located on the front brake plate. The brake should be adjusted so that the wheel is free to revolve before pressure is applied to the handlebar lever and is applied fully before the handlebar lever touches the handlebar. Make sure the adjuster locknut is tight after the correct adjustment has been made.

2 The rear brake is adjusted by means of the adjusting nut on the end of the brake operating rod. This nut is self-locking. Adjustment is largely a matter of personal choice, but excessive travel of the footbrake pedal should not be necessary before the brake is applied fully.

3 Check frequently that the rear brake torque arm bolt is right (all swinging arm models). If the torque arm becomes detached, the rear brake will lock in the full-on position immediately it is applied and may give rise to a serious accident.

4 Efficient brakes depend on good leverage of the operating arms. The angle between the brake operating arm and the cable or rod should never exceed 90° when the brake is fully applied.

10 Rear Wheel Sprocket - Removal, Examination and Replacement

1 The rear wheel sprocket is attached to the hub flange by eight bolts, each fitted with spring washers. If these bolts are removed, the sprocket can be detached.

2 Check the condition of the sprocket teeth. If they are hooked chipped or badly worn, the sprocket should be replaced.

3 It is bad practice to renew one sprocket on its own. The final drive sprockets should always be renewed as a pair and a new chain fitted, otherwise rapid wear will necessitate even earlier replacement.

11 Final Drive Chain - Examination and Lubrication

1 The final drive chain does not have the benefit of full enclosure or positive lubrication, in which case it will require attention from time to time, particularly when the machine is used on wet or dirty roads.

2 Chain adjustment is correct when there is approximately ¾ in play in the middle of the run on the rigid frame models and the swinging arm models, or ½ in play on the plunger sprung models. Always check at the tightest spot on the chain run under load.

3 If the chain is too slack, adjustment is effected by slackening the rear wheel nuts and also the torque arm bolt (where fitted),

then drawing the wheel backwards by means of the cycle-type drawbolt adjusters at the frame ends. Make sure each adjuster is turned an equal amount, so that the rear wheel is kept centrally-disposed within the frame. When the correct adjusting point has been reached, push the wheel hard forward, then tighten the wheel nuts, not forgetting the torque arm bolt, if fitted. Re-check the chain tension and also the wheel alignment before the final tightening of all nuts.

4 Application of engine oil to the chain from time to time will serve as a satisfactory form of lubrication, but it is preferable to remove the chain every 2,000 miles and clean it in a bath of paraffin before immersing it in a special chain lubricant such as "Linklyfe". This latter tyoe of lubricant achieves better and more lasting penetration of the chain links and rollers and is less likely to be thrown off when the chain is in motion.

5 To check whether the chain needs replacing lay it lengthwise in a straight line and compress it, so that all play is taken up. Anchor one end and then pull on the other, to stretch the chain in the opposite direction. If the chain extends by more than the distance between two adjacent rollers, replacement is advised.

6 When replacing the chain, make sure the spring link is positioned correctly, with the closed end facing the direction of travel. Reconnection is made easier if the ends of the chain are pressed into the teeth of the rear wheel sprocket.

12 Rear Wheel - Replacement

1 The rear wheel is replaced in the frame by reversing the dismantling procedure given in Section 6. Check that the speedometer drive gearbox is positioned correctly on the rear wheel spindle so that the driving dogs engage with the hub and also that the distance piece is in place. When the wheel is in position, check that the speedometer drive revolves before connecting the speedometer drive cable, then retension the final drive chain before tightening the wheel nuts. Do not overlook the rear brake torque arm fixing, if the machine has the later type of brake plate.

13 Tyres - Removal and Replacement

1 At some time or other the need will arise to remove and replace the tyres, either as the result of a puncture or because a replacement is required to offset wear. To the inexperienced, tyre changing represents a formidable task yet if a few simple rules are observed and the technique learned the whole operation is surprisingly simple.

2 To remove the tyre from either wheel, first detach the wheel from the machine by following the procedure in Chapters 4.2 or 5.6, depending on whether the front or the rear wheel is involved. Deflate the tyre by removing the valve insert and when it is fully deflated, push the bead of the tyre away from the wheel rim on both sides so that the bead enters the centre well of the rim. Remove the locking cap and push the tyre valve into the tyre itself.

3 Insert a tyre lever close to the valve and lever the edge of the tyre over the outside of the wheel rim. Very little force should be necessary; if resistance is encountered it is probably due to the fact that the tyre beads have not entered the well of the wheel rim all the way round the tyre.

4 Once the tyre has been edged over the wheel rim, it is easy to work around the wheel rim so that the tyre is completely free on one side. At this stage, the inner tube can be removed.

5 Working from the other side of the wheel, ease the other edge of the tyre over the outside of the wheel rim that is furthest away. Continue to work around the rim until the tyre is free completely from the rim.

6 If a puncture has necessitated the removal of the tyre, re-inflate the inner tube and immerse it in a bowl of water to trace the source of the leak. Mark its position and deflate the tube. Dry the tube and clean the area around the puncture with a petrol-soaked rag. When the surface has dried, apply the rubber

7.1. Rear brake assembly is identical to front

8.1A. Pack hub with new grease

8.1B. Grease bearings liberally

10.2. Badly worn sprocket teeth

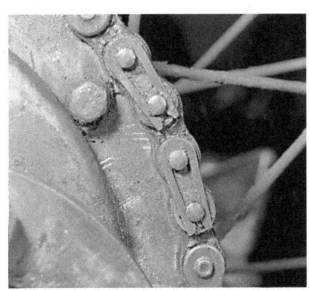

11.6. Closed end of spring link faces direction of travel (It is bad practice to fit two links to lengthen chain)

15.1. Interior of speedometer drive gearbox must be well greased

solution and allow this to dry before removing the backing from the patch and applying the patch to the surface.

7 It is best to use a patch of the self-vulcanising type, which will form a very permanent repair. Note that it may be necessary to remove a protective covering from the top surface of the patch, after it has sealed in position. Inner tubes made from synthetic rubber may require a special type of patch and adhesive, if a satisfactory bond is to be achieved.

8 Before replacing the tyre, check the inside to make sure the agent that caused the puncture is not trapped. Check also the outside of the tyre, particularly the tread area, to make sure nothing is trapped that may cause a further puncture.

9 If the inner tube has been patched on a number of past occasions, or if there is a tear or large hole, it is preferable to discard it and fit a replacement. Sudden deflation may cause an accident, particularly if it occurs with the front wheel.

10 To replace the tyre, inflate the inner tube sufficiently for it to assume a circular shape but only just. Then push it into the tyre so that it is enclosed completely. Lay the tyre on the wheel at an angle and insert the valve through the rim tape and the hole in the wheel rim. Attach the locking cap on the first few threads, sufficient to hold the valve captive in its correct location.

11 Starting at the point furthest from the valve, push the tyre bead over the edge of the wheel rim until it is located in the central well. Continue to work around the tyre in this fashion until the whole of one side of the tyre is on the rim. It may be necessary to use a tyre lever during the final stages.

12 Make sure there is no pull on the tyre valve and again commencing with the area furthest from the valve, ease the other bead of the tyre over the edge of the rim. Finish with the area close to the valve, pushing the valve up into the tyre until the locking cap touches the rim. This will ensure the inner tube is not trapped when the last section of the bead is edged over the rim with a tyre lever.

13 Check that the inner tube is not trapped at any point. Re-inflate the inner tube, and check that the tyre is seating correctly around the wheel rim. There should be a thin rib moulded around the wall of the tyre on both sides, which should be equidistant from the wheel rim at all points. If the tyre is unevenly located on the rim, try bouncing the wheel when the tyre is at the recommended pressure. It is probable that one of the beads has not pulled clear of the centre well.

14 Always run the tyres at the recommended pressures and never under or over-inflate. The correct pressures for solo use are 17 psi front and 22 psi rear. If a pillion passenger is carried, increase the rear tyre pressure only to 27 psi.

15 Tyre replacement is aided by dusting the side walls, particularly in the vicinity of the beads, with a liberal coating of french chalk. Washing-up liquid can also be used to good effect, but this has the disadvantage of causing the inner surfaces of the wheel rim to rust.

16 Never replace the inner tube and tyre without the rim tape in position. If this precaution is overlooked there is good chance of the ends of the spoke nipples chafing the inner tube and causing a crop of punctures.

17 Never fit a tyre that has a damaged tread or side walls. Apart from the legal aspects, there is a very great risk of a blow-out, which can have serious consequences on any two-wheel vehicle.

18 Tyre valves rarely give trouble, but it is always advisable to check whether the valve itself is leaking before removing the tyre. Do not forget to fit the dust cap, which forms an effective second seal.

14 Security Bolts (Competition models only)

1 It is often necessary to run the tyres fitted to the Competition models at low pressures, in order to obtain the benefit of greatly improved wheel grip on rough terrain. Under these circumstances, the tyre tends to creep on the wheel rim unless it can be restrained in some way. The security bolt fulfills this requirement in a simple and effective manner, to prevent the valve being torn from the inner tube body as it is dragged with the outer cover.

2 A security bolt is fitted to each wheel of the "Bushman" models and in the case of the early Competition models, usually to the rear wheel only. Before attempting to remove or replace a tyre, the security bolt must be slackened off completely because it clamps the bead of the tyre to the wheel rim.

15 Speedometer Drive Gearbox - General

1 It is unlikely that the speedometer drive gearbox will require attention during the normal service life of the machine, provided it has been greased at regular intervals.

2 Speedometer drive gearboxes are not necessarily interchangeable and it is advisable to check when a replacement has to be fitted. The drive ratio is related to the size of the rear wheel and also the section of the tyre fitted, two variables that will have a marked effect on the accuracy of the speedometer reading.

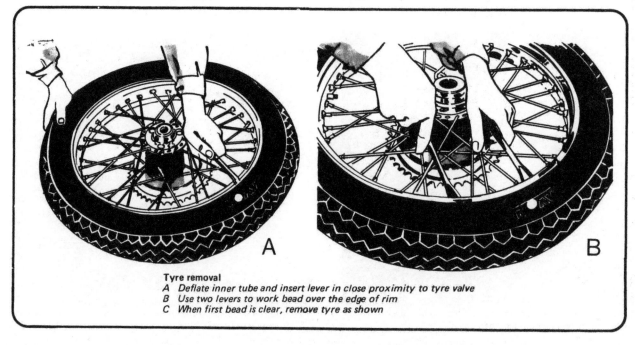

Tyre removal
A Deflate inner tube and insert lever in close proximity to tyre valve
B Use two levers to work bead over the edge of rim
C When first bead is clear, remove tyre as shown

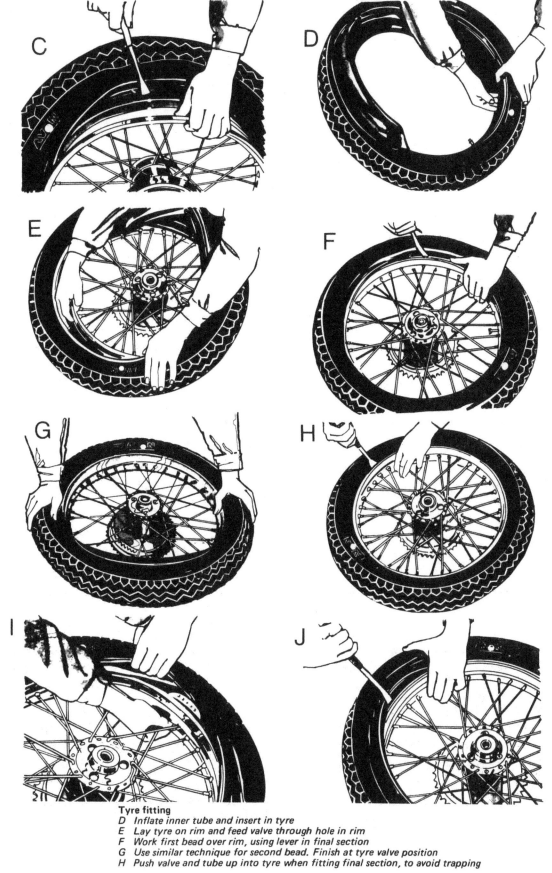

Tyre fitting
D Inflate inner tube and insert in tyre
E Lay tyre on rim and feed valve through hole in rim
F Work first bead over rim, using lever in final section
G Use similar technique for second bead. Finish at tyre valve position
H Push valve and tube up into tyre when fitting final section, to avoid trapping

Security bolts (Competition models)
I Fit the security bolt very loosely when one bead of the tyre is fitted
J Then fit tyre in normal way. Tighten bolt when tyre is properly seated

16 Fault Diagnosis - Wheels, Brakes and Final Drive

Symptom	Cause	Remedy
Handlebars oscillate at low speeds	Buckle or flat in wheel rim, most probably front wheel	Check rim alignment by spinning wheel. Correct by retensioning spokes or by having wheel rebuilt on new rim.
	Tyre not straight on rim	Check tyre alignment.
Machine lacks power and accelerates poorly	Brakes binding	Warm brake drums provide best evidence. Re-adjust brakes.
Brakes grab when applied gently	Ends of brake shoes not chamfered	Chamfer with file.
	Elliptical brake drum	Lightly skim in lathe (specialist attention needed).
Brake pull-off sluggish	Brake cam binding in housing	Free and grease.
	Weak brake shoe springs	Replace if springs not displaced.
Harsh transmission	Worn or badly adjusted chains	Adjust or replace as necessary
	Hooked or badly worn sprockets	Replace as a pair.

Chapter 6 Electrical equipment

Contents

Specifications

The general specification and history of the various types of lighting systems fitted since the inception of the 'Bantam' model in 1948 is given in the introduction to Chapter 3 of this manual. Data relating to the ancillary equipment is as follows:

Battery (lead acid, 6 volt)

Make	Lucas (1948 - 1953)
	Varley (1954 - 1955)
	Lucas (1956 onwards)
Capacity	5 amp hr (1948 - 1953)
	9 amp hr (1954 - 1955)
	11 amp hr (1956 onwards)
Earth connection	Positive earth

All models fitted with Wico-Pacy direct lighting have no battery in the generally accepted sense. A type 300 cycle lamp battery is housed within the headlamp shell to provide facilities for parking lights only. This battery cannot be recharged and is therefore expendable.

Lighting

Headlmap bulb	24/24W (Wico-Pacy direct lighting)
	30/24W (Early Lucas battery lighting set)
	30/24W Pre-focus (Wico-Pacy and Lucas 60W alternators)
	24/24W Pre-focus ('Bushman' direct lighting)
Pilot lamp bulb	0.2A, 2.5v (Wico-Pacy direct lighting)
	3W, 6 volt (Lucas and Wico-Pacy battery lighting sets)
Rear/stop lamp bulb	0.5A, 6v (Wico-Pacy direct lighting)
	3W, 6 volt (Early Lucas battery lighting set)
	6/18W (Lucas and Wico-Pacy 60W alternators)

Present day statutory requirements have dictated the need for a combined stop and rear lamp, the latter of which must have a minimum rating of 6 watts. The 60 watt alternators have the capacity to meet the additional electrical load, without imposing any drain on the battery.

1 General Description

1 Two types of lighting system are fitted to the "Bantam" models, the choice depending on the specification of the machine. The early D1 models employ a direct lighting system, powered by lighting coils in the flywheel magneto generator. The absense of a battery means the lights can be used only when the engine is running and it is necessary to include a cycle lamp dry battery in the circuit so that low voltage bulbs can be brought into use for parking.

2 A De Luxe version of the model D1 was introduced for those who required a better lighting system. This variant uses an a.c. alternator and separate ignition coil, in place of the flywheel magneto generator. The alternator provides a higher output, which is converted into direct current by a rectifier. With a ready-available source of direct current, it is now possible to charge a battery, which can be included in the circuit and used to run the various electrical components. A higher electrical load can be tolerated and maintained without imposing a drain on the battery, a situation that encourages the use of the much improved lighting system.

3 Progress in the electrical industry has led to the use of even higher output generators, such as the Wico-Pacy type 1G 1768 fitted to the D14 and "Bushman" models at the time when

"Bantam" production ceased. A particularly interesting feature is the fact that the "Bushman" model uses a direct lighting system with energy transfer facilities, yet has a lighting set capable of matching existing battery-operated systems, with none of the earlier disadvantages.

2 Flywheel Generators - Checking Output

As explained in Chapter 3, the electrical output from any of the generators fitted to the "Bantam" models can be checked only with specialised equipment of the multi-meter type. If the generator is suspect, it should be checked by either a BSA agent or an auto-electrical expert.

3 Battery - Examination and Maintenance

1 Batteries of different amp. hour capacities have been fitted to the various "Bantam" models, the capacity of the battery being dependent on the current rating of the full electrical load. A battery should always be replaced with another of similar capacity; it is false economy to fit a battery having a lower capacity because the cost advantage will soon be offset by the reduced working life.
2 Maintenance is normally limited to keeping the electrolyte level just above the plates and separators. Modern batteries have translucent plastics cases, which make the check of electrolyte level much easier.
3 Unless acid is spilt, which may occur if the machine falls over, the electrolyte should always be topped up with distilled water until the correct level is restored. If acid is spilt on any part of the machine, it should be neutralised with an alkali such as washing soda and washed away with plenty of water, otherwise serious corrosion will occur. Top up with sulphuric acid of the correct specific gravity (1.260 - 1.280) ONLY when spillage has occurred.
4 It is seldom practicable to repair a cracked battery case because the acid already in the joint will prevent the formation of an effective seal. It is always best to replace a cracked battery, especially in view of the corrosion that will be caused by the leakage of acid.
5 The Varley battery, fitted to the 1954 and 1955 models is of the lead/acid type, but has the electrolyte absorbed in glass wool packing that surrounds the plates. This obviates the risk of acid spillage when the battery is inverted. Topping up consists of adding a teaspoonful of distilled water to each cell and leaving the battery to stand for approximately five minutes before draining off any excess.

4 Battery - Charging Procedure

1 Whilst the machine is running, it is unlikely that the battery will require attention other than routine maintenance because the generator will keep it charged. However, if the machine is used for a succession of short journeys only, mainly during the hours of darkness when the lights are in full use, it is possible that the output from the generator will be able to keep pace with the heavy electrical demand, especially on the earlier models. Under these circumstances it will be necessary to remove the battery from time to time, to have it charged independently.
2 The normal charging rate is 1 amp. A more rapid charge can be given in an emergency, but this should be avoided if possible because it will shorten the working life of the battery.
3 When the battery has been removed from a machine that has been laid up, a 'refresher' charge should be given every six weeks if the battery is to be maintained in good condition.

5 Selenium Rectifier - General Description

1 The function of the selenium rectifier is to convert the a.c. produced by the generator to d.c. so that it can be used to charge the battery and operate the lighting circuit etc. The usual symptom of a defective rectifier is a battery that discharges rapidly because it is receiving no charge from the generator.
2 The rectifier is located where it is not exposed to water or battery acid, which will cause it to malfunction. The question of access is of relatively little importance because the rectifier is unlikely to give trouble during normal operating conditions. It is not practicable to repair a damaged rectifier; replacement is the only satisfactory solution. One of the most frequent causes of rectifier failure is the inadvertent connection of the battery in reverse, which results in a reverse current flow.
3 Damage to the rectifier is also liable to occur if the machine is run without a battery for any period of time (battery lighting models only). A high voltage will develop in the absence of any load on the coil, which will cause a reverse flow of current and consequent damage to the rectifier cells.
4 It is not possible to check whether the rectifier is functioning correctly without the appropriate test equipment. A BSA agent or an auto-electrical expert are best qualified to advise in such cases.
5 Do not loosen the rectifier locking nut or bend, cut, scratch or rotate the selenium wafers. Any such action will cause the electrode alloy coating to peel and destroy the working action.

6 Headlamp - Replacing Bulbs and Adjusting Beam Height

1 To remove the headlamp rim, unscrew the screw at the base of the rim. The rim will then pull off, complete with the reflector unit and bulbs.
2 The reflector unit contains a double-filament bulb that provides the main and dipped headlamp beams, and a pilot lamp bulb, for parking use. The double-filament headlamp bulb is controlled from a dipswitch mounted on the handlebars or, in the case of some of the earlier models, from a headlamp lighting switch mounted on the right-hand side of the handlebars. The pilot lamp is found above the main bulb (early direct lighting models), attached to the main bulb holder (1950-1 battery lighting models) or in an underslung housing immediately below the main headlamp shell (1952-3 battery lighting models). Thereafter it was positioned below the main headlamp bulb to protrude through the reflector. Note that the "Bushman" models do not have a pilot lamp.
3 It is not necessary to refocus the headlamp when a new bulb is fitted because the bulbs are of the prefocus type. To release the bulb holder from the reflector, twist and pull.
4 Beam height is adjusted by slackening the two headlamp shell retaining bolts and tilting the headlamp either upwards or downwards. Adjustments should always be made with the rider normally seated.
5 UK lighting regulations stipulate that the lighting system must be arranged so that the light will not dazzle a person standing in the same horizontal plane as the vehicle at a distance greater than 25 yards from the lamp, whose eye level is not less than 3 feet 6 inches above that plane. It is easy to approximate this setting by placing the machine 25 yards away from a wall, on a level road, and setting the beam height so that it is concentrated at the same height as the distance from the centre of the headlamp to the ground. The rider must be seated normally during this operation and also the pillion passenger, if one is carried regularly.

7 Tail and Stop Lamp - Replacing Bulbs

1 All except the early models have a double filament bulb of 6/18W rating, to illuminate the rear of the machine and the rear number plate, and to give visual warning when the rear brake is

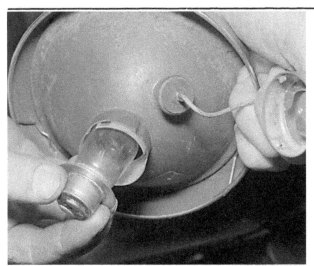

6.2A. Prefocus bulb has flange to determine correct mounting position

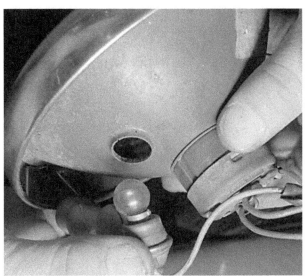

6.2B. Pilot bulb holder is push fit in reflector

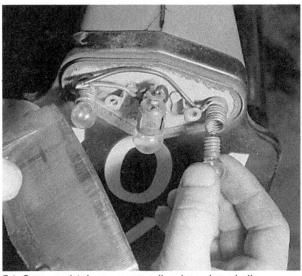

7.1. Some models have separate tail and stop lamp bulbs

applied. To gain access to the bulb, unscrew the two screws that retain the moulded plastics lens cover to the rear lamp assembly and remove the cover and sealing gasket.

2　Early models (pre-1954) are fitted with a combined tail and stop lamp only if the machine has battery lighting. Access to the bulb fitted to these models and also to the rear lamp bulb of the direct lighting models is accomplished by turning the rear half of the lamp cover in an anti-clockwise direction, to release it from the bayonet catch.

3　The stop lamp is operated by a stop lamp switch on the left-hand side of the machine, operated by the brake rod via a spring. The switch does not require any special attention apart from the occasional application of thin oil.

8　Speedometer Bulb - Replacement

1　Some models are fitted with an internally-illuminated speedometer head. The bulb holder screws into the base of the speedometer head and is fitted with a 6 volt bulb rated at 0.6W.

9　Horn - Location and Adjustment

1　The early models have a bulb horn that passes through the centre of the steering head. No attention is required unless the horn loses its voice, in which case the reed will have to be replaced.

2　Later models have an electrically-operated horn of the electro-magnetic type, operated from a push button on the handlebars. The horn is mounted below the nose of the dual seat, in some cases within the left-hand tol box cover.

3　A small serrated screw located near the terminals affords a means of adjustment. If the horn does not function, turn this screw anti-clockwise until the restored note just fails to sound and then turn it back clockwise about one quarter of a turn.

10　Wiring - Layout and Inspection

1　The wiring harness is colour-coded and will correspond with the accompanying wiring diagrams.

2　Visual inspection will show whether any breaks or frayed outer coverings are giving rise to short circuits. Another source of trouble may be the snap connectors, particularly where the connector has not been pushed home fully in the outer casing.

3　Intermittent short circuits can often be traced to a chafed wire that passes through or close to a metal component, such as a frame member. Avoid tight bends in the wire or situations where the wire can become trapped or stretched between casings.

11　Ignition and Lighting Switches

1　Various types of lighting switch have been fitted since the inception of the "Bantam", some of which contain the ignition switch and depend on the insertion of the ignition key before they can be operated. The key cannot be removed whilst the ignition is switched on.

2　The remote headlamp switch, fitted to the early models with Wico-Pacy direct lighting, clamps to the handlebars in a manner similar to the conventional handlebar controls. Control is effected by means of a Bowden cable, which has an adjuster for synchronising the control sequence with the inscriptions on the switch top. Switch action is controlled by a simple friction damper in the top of the switch, actuated by a screw.

3　The other types of switch, particularly those that contain provision for the ignition circuit, are virtually impossible to repair if they prove defective. A new replacement should be fitted if the switch gives trouble, rather than risk sudden failure of either the ignition or the lights as the result of a bad or intermittent contact.

4　On no account oil the switch or the oil will spread across the internal contacts and form an effective insulator.

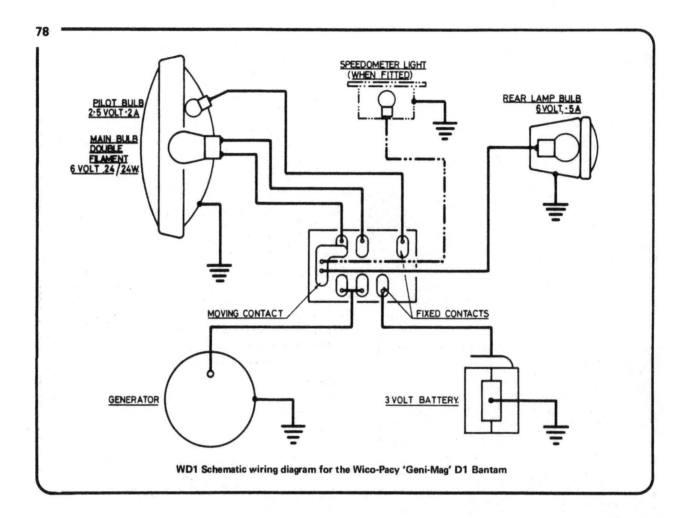

WD1 Schematic wiring diagram for the Wico-Pacy 'Geni-Mag' D1 Bantam

KEY TO CABLE COLOURS

1	BLUE	33	BROWN
2	BLUE with RED	34	BROWN with RED
3	BLUE with YELLOW	35	BROWN with YELLOW
4	BLUE with WHITE	36	BROWN with BLUE
5	BLUE with GREEN	37	BROWN with WHITE
6	BLUE with PURPLE	38	BROWN with GREEN
7	BLUE with BROWN	39	BROWN with PURPLE
8	BLUE with BLACK	40	BROWN with BLACK
9	WHITE	41	RED
10	WHITE with RED	42	RED with YELLOW
11	WHITE with YELLOW	43	RED with BLUE
12	WHITE with BLUE	44	RED with WHITE
13	WHITE with GREEN	45	RED with GREEN
14	WHITE with PURPLE	46	RED with PURPLE
15	WHITE with BROWN	47	RED with BROWN
16	WHITE with BLACK	48	RED with BLACK
17	GREEN	49	PURPLE
18	GREEN with RED	50	PURPLE with RED
19	GREEN with YELLOW	51	PURPLE with YELLOW
20	GREEN with BLUE	52	PURPLE with BLUE
21	GREEN with WHITE	53	PURPLE with WHITE
22	GREEN with PURPLE	54	PURPLE with GREEN
23	GREEN with BROWN	55	PURPLE with BROWN
24	GREEN with BLACK	56	PURPLE with BLACK
25	YELLOW	57	BLACK
26	YELLOW with RED	58	BLACK with RED
27	YELLOW with BLUE	59	BLACK with YELLOW
28	YELLOW with WHITE	60	BLACK with BLUE
29	YELLOW with GREEN	61	BLACK with WHITE
30	YELLOW with PURPLE	62	BLACK with GREEN
31	YELLOW with BROWN	63	BLACK with PURPLE
32	YELLOW with BLACK	64	BLACK with BROWN

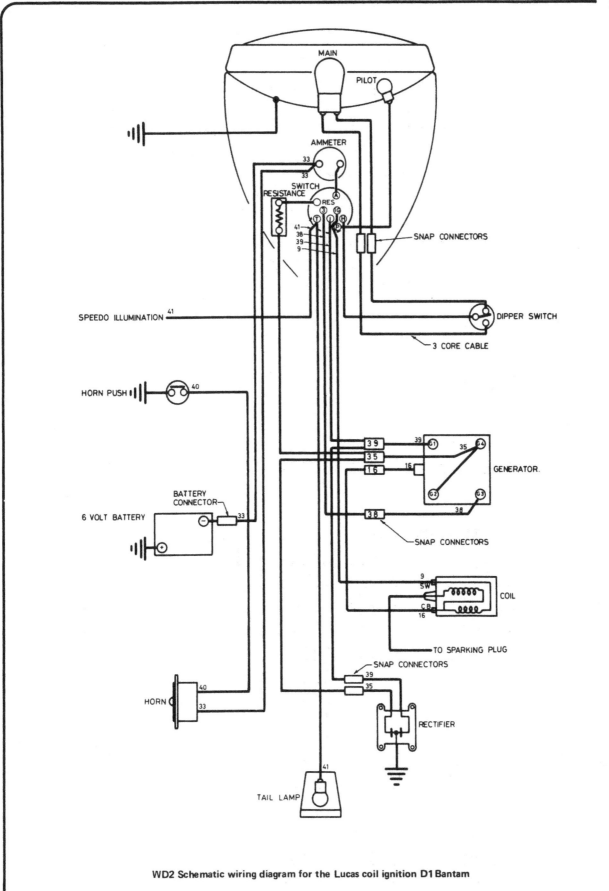

WD2 Schematic wiring diagram for the Lucas coil ignition D1 Bantam

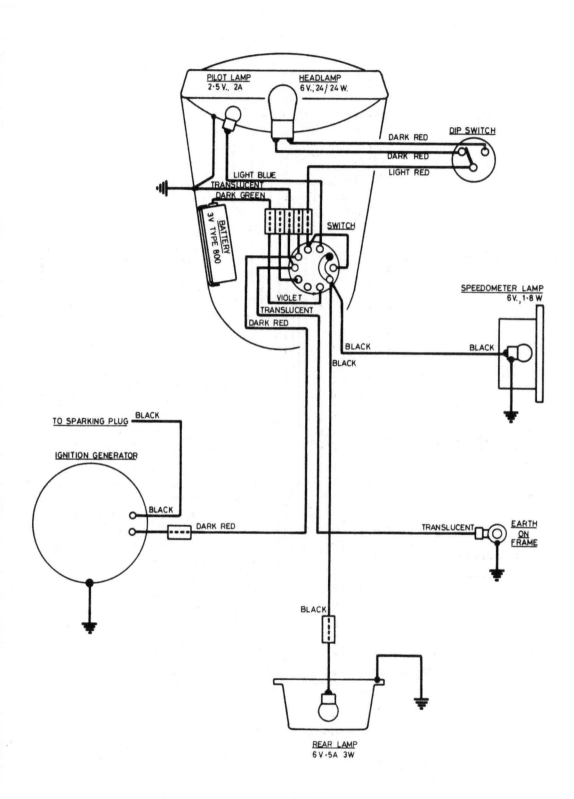

WD3 Schematic wiring diagram for 'direct lighting' Wico-Pacy Series 55/Mark 8 equipped D1, D3, D5 and D7 Bantams

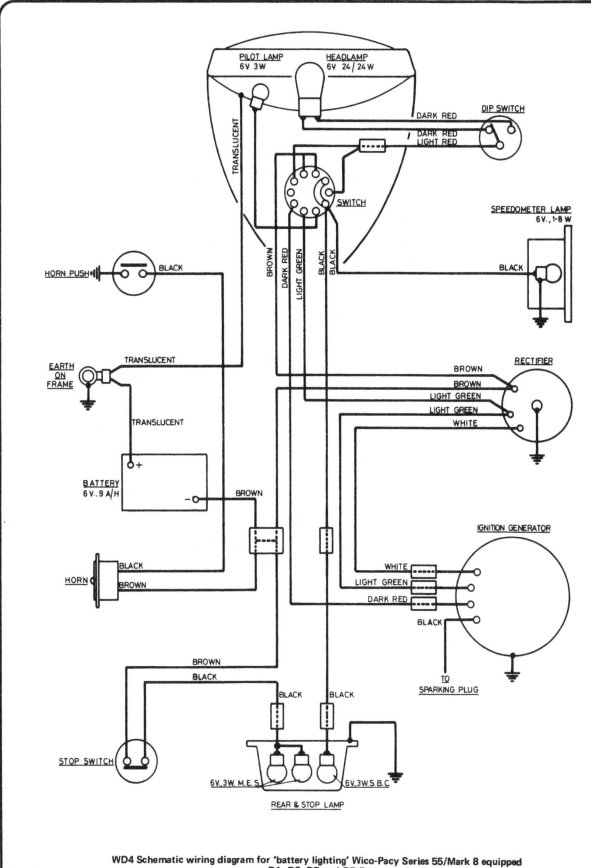

WD4 Schematic wiring diagram for 'battery lighting' Wico-Pacy Series 55/Mark 8 equipped D1, D3, D5 and D7 Bantams

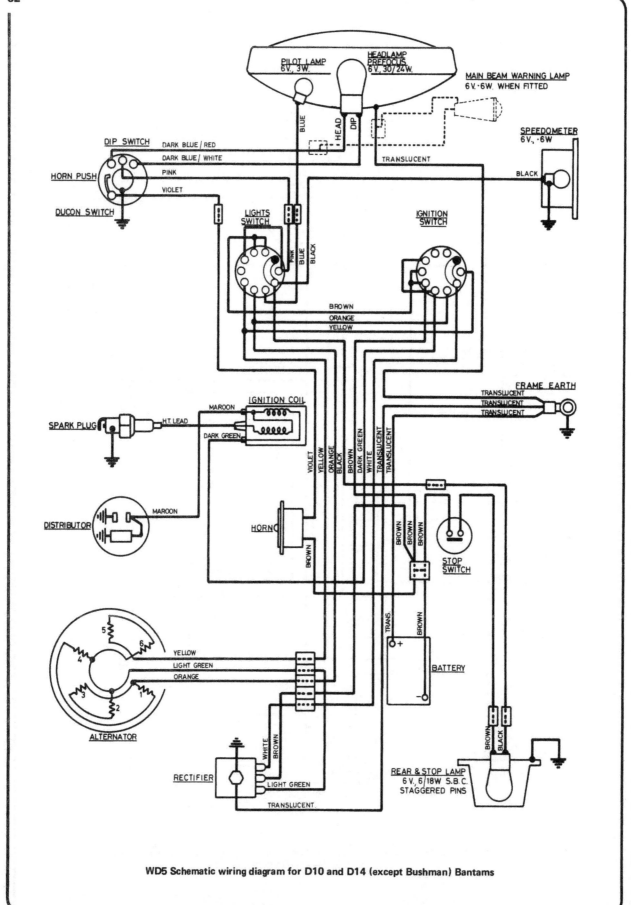

WD5 Schematic wiring diagram for D10 and D14 (except Bushman) Bantams

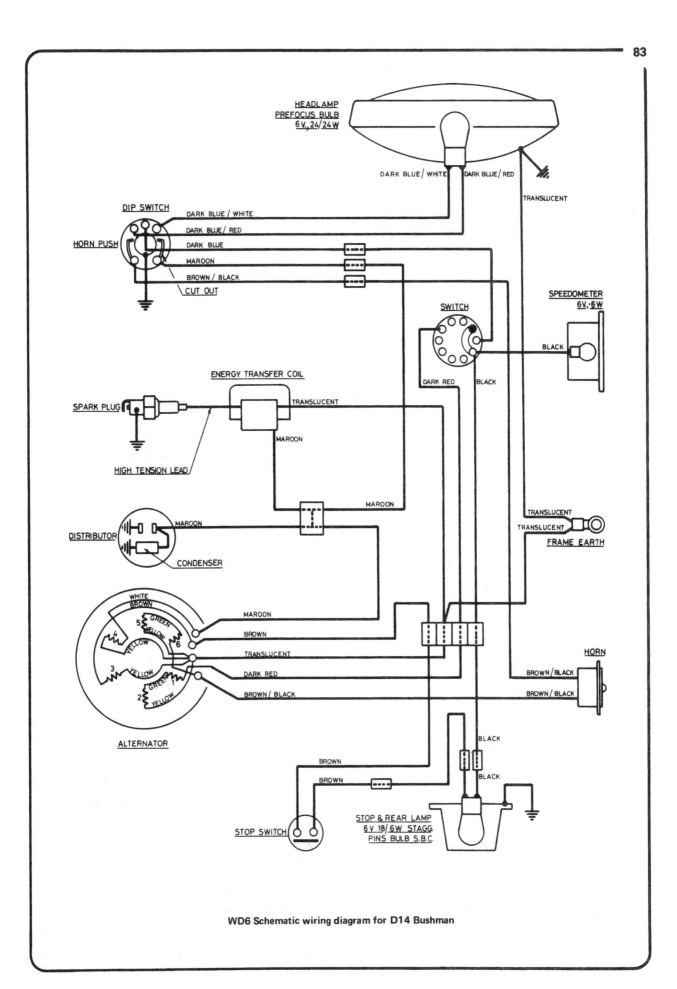

WD6 Schematic wiring diagram for D14 Bushman

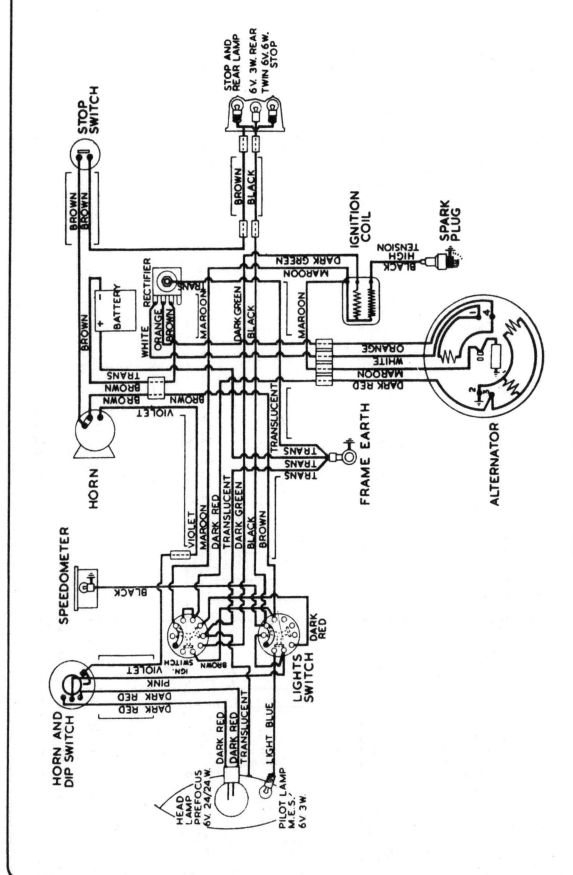

Wiring diagram for D7 model – two switch type

Metric conversion tables

Inches	Decimals	Millimetres	Millimetres to Inches		Inches to Millimetres	
			mm	Inches	Inches	mm
1/64	0.015625	0.3969	0.01	0.00039	0.001	0.0254
1/32	0.03125	0.7937	0.02	0.00079	0.002	0.0508
3/64	0.046875	1.1906	0.03	0.00118	0.003	0.0762
1/16	0.0625	1.5875	0.04	0.00157	0.004	0.1016
5/64	0.078125	1.9844	0.05	0.00197	0.005	0.1270
3/32	0.09375	2.3812	0.06	0.00236	0.006	0.1524
7/64	0.109375	2.7781	0.07	0.00276	0.007	0.1778
1/8	0.125	3.1750	0.08	0.00315	0.008	0.2032
9/64	0.140625	3.5719	0.09	0.00354	0.009	0.2286
5/32	0.15625	3.9687	0.1	0.00394	0.01	0.254
11/64	0.171875	4.3656	0.2	0.00787	0.02	0.508
3/16	0.1875	4.7625	0.3	0.01181	0.03	0.762
13/64	0.203125	5.1594	0.4	0.01575	0.04	1.016
7/32	0.21875	5.5562	0.5	0.01969	0.05	1.270
15/64	0.234375	5.9531	0.6	0.02362	0.06	1.524
1/4	0.25	6.3500	0.7	0.02756	0.07	1.778
17/64	0.265625	6.7469	0.8	0.03150	0.08	2.032
9/32	0.28125	7.1437	0.9	0.03543	0.09	2.286
19/64	0.296875	7.5406	1	0.03937	0.1	2.54
5/16	0.3125	7.9375	2	0.07874	0.2	5.08
21/64	0.328125	8.3344	3	0.11811	0.3	7.62
11/32	0.34375	8.7312	4	0.15748	0.4	10.16
23/64	0.359375	9.1281	5	0.19685	0.5	12.70
3/8	0.375	9.5250	6	0.23622	0.6	15.24
25/64	0.390625	9.9219	7	0.27559	0.7	17.78
13/32	0.40625	10.3187	8	0.31496	0.8	20.32
27/64	0.421875	10.7156	9	0.35433	0.9	22.86
7/16	0.4375	11.1125	10	0.39370	1	25.4
29/64	0.453125	11.5094	11	0.43307	2	50.8
15/32	0.46875	11.9062	12	0.47244	3	76.2
31/64	0.484375	12.3031	13	0.51181	4	101.6
1/2	0.5	12.7000	14	0.55118	5	127.0
33/64	0.515625	13.0969	15	0.59055	6	152.4
17/32	0.53125	13.4937	16	0.62992	7	177.8
35/64	0.546875	13.8906	17	0.66929	8	203.2
9/16	0.5625	14.2875	18	0.70866	9	228.6
37/64	0.578125	14.6844	19	0.74803	10	254.0
19/32	0.59375	15.0812	20	0.78740	11	279.4
39/64	0.609375	15.4781	21	0.82677	12	304.8
5/8	0.625	15.8750	22	0.86614	13	330.2
41/64	0.640625	16.2719	23	0.90551	14	355.6
21/32	0.65625	16.6687	24	0.94488	15	381.0
43/64	0.671875	17.0656	25	0.98425	16	406.4
11/16	0.6875	17.4625	26	1.02362	17	431.8
45/64	0.703125	17.8594	27	1.06299	18	457.2
23/32	0.71875	18.2562	28	1.10236	19	482.6
47/64	0.734375	18.6531	29	1.14173	20	508.0
3/4	0.75	19.0500	30	1.18110	21	533.4
49/64	0.765625	19.4469	31	1.22047	22	558.8
25/32	0.78125	19.8437	32	1.25984	23	584.2
51/64	0.796875	20.2406	33	1.29921	24	609.6
13/16	0.8125	20.6375	34	1.33858	25	635.0
53/64	0.828125	21.0344	35	1.37795	26	660.4
27/32	0.84375	21.4312	36	1.41732	27	685.8
55/64	0.859375	21.8281	37	1.4567	28	711.2
7/8	0.875	22.2250	38	1.4961	29	736.6
57/64	0.890625	22.6219	39	1.5354	30	762.0
29/32	0.90625	23.0187	40	1.5748	31	787.4
59/64	0.921875	23.4156	41	1.6142	32	812.8
15/16	0.9375	23.8125	42	1.6535	33	838.2
61/64	0.953125	24.2094	43	1.6929	34	863.6
31/32	0.96875	24.6062	44	1.7323	35	889.0
63/64	0.984375	25.0031	45	1.7717	36	914.4

English/American terminology

Because this book has been written in England, British English component names, phrases and spellings have been used throughout. American English usage is quite often different and whereas normally no confusion should occur, a list of equivalent terminology is given below.

English	American	English	American
Air filter	Air cleaner	Number plate	License plate
Alignment (headlamp)	Aim	Output or layshaft	Countershaft
Allen screw/key	Socket screw/wrench	Panniers	Side cases
Anticlockwise	Counterclockwise	Paraffin	Kerosene
Bottom/top gear	Low/high gear	Petrol	Gasoline
Bottom/top yoke	Bottom/top triple clamp	Petrol/fuel tank	Gas tank
Bush	Bushing	Pinking	Pinging
Carburettor	Carburetor	Rear suspension unit	Rear shock absorber
Catch	Latch	Rocker cover	Valve cover
Circlip	Snap ring	Selector	Shifter
Clutch drum	Clutch housing	Self-locking pliers	Vise-grips
Dip switch	Dimmer switch	Side or parking lamp	Parking or auxiliary light
Disulphide	Disulfide	Side or prop stand	Kick stand
Dynamo	DC generator	Silencer	Muffler
Earth	Ground	Spanner	Wrench
End float	End play	Split pin	Cotter pin
Engineer's blue	Machinist's dye	Stanchion	Tube
Exhaust pipe	Header	Sulphuric	Sulfuric
Fault diagnosis	Trouble shooting	Sump	Oil pan
Float chamber	Float bowl	Swinging arm	Swingarm
Footrest	Footpeg	Tab washer	Lock washer
Fuel/petrol tap	Petcock	Top box	Trunk
Gaiter	Boot	Torch	Flashlight
Gearbox	Transmission	Two/four stroke	Two/four cycle
Gearchange	Shift	Tyre	Tire
Gudgeon pin	Wrist/piston pin	Valve collar	Valve retainer
Indicator	Turn signal	Valve collets	Valve cotters
Inlet	Intake	Vice	Vise
Input shaft or mainshaft	Mainshaft	Wheel spindle	Axle
Kickstart	Kickstarter	White spirit	Stoddard solvent
Lower leg	Slider	Windscreen	Windshield
Mudguard	Fender		

Index